cSUR-UT Series: Library for Sustainable Urban Regeneration
Volume 7

Series Editor: Shinichiro Ohgaki, Tokyo, Japan

cSUR-UT Series: Library for Sustainable Urban Regeneration

By the process of urban development in the 20th century, characterized by suburban expansion and urban redevelopment, many huge and sophisticated complexes of urban structures have been erected in developed countries. However, with conventional technologies focused on the construction of structures, it has become difficult to keep urban spaces adaptable to environmental constraints and economic, social and cultural changes. In other words, it has become difficult for conventional technologies to meet social demands for the upgrading of social capital in a sustainable manner and for the regeneration of attractive urban space that is not only safe and highly efficient but also conscious of historical, cultural and local identities to guarantee a high quality of life for all. Therefore, what is needed now is the creation of a new discipline that is able to reorganize the existing social capital and the technologies to implement it.

For this purpose, there is a need to go beyond the boundaries of conventional technologies of construction and structural design and to integrate the following technologies:

(1) Technology concerned with environmental and risk management

(2) Technology of conservation and regeneration with due consideration to the local characteristics of existing structures including historical and cultural resources

(3) Technologies of communication, consensus building, plan making and space management to coordinate and integrate the individual activities initiated by various actors of society

Up to now, architecture, civil engineering, and urban engineering in their respective fields have, while dealing with different time-space scales and structures, accumulated cutting-edge knowledge and contributed to the formation of favorable urban spaces. In the past, when emphasis was put on developing new residential areas and constructing new structures, development and advancement of such specialized disciplines were found to be the most effective.

However, current problems confronting urban development can be highlighted by the fact that a set of optimum solutions drawn from the best practices of each discipline is not necessarily the best solution. This is especially true where there are relationships of trade-offs among such issues as human risk and environmental load. In this way, the integration of the above three disciplines is strongly called for.

In order to create new integrated knowledge for sustainable urban regeneration, the Center for Sustainable Urban Regeneration (cSUR), The University of Tokyo, was established in 2003 as a core organization of one of the 21st Century Centers of Excellence Programs funded by the Ministry of Education and Science, Japan, and cSUR has coordinated international research alliances and collaboratively engages with common issues of sustainable urban regeneration.

The cSUR series are edited and published to present the achievements of our collaborative research and new integrated approaches toward sustainable urban regeneration.

T. Kidokoro, N. Harata, L. P. Subanu,
J. Jessen, A. Motte, E. P. Seltzer (Eds.)

Sustainable City Regions:

Space, Place and Governance

Springer

Tetsuo Kidokoro, Ph.D.
Associate Professor
Department of Urban Engineering
The University of Tokyo
7-3-1 Hongo, Bunkyo-ku
Tokyo 113-8656, Japan

Noboru Harata, Ph.D.
Professor
Department of Urban Engineering
The University of Tokyo
7-3-1 Hongo, Bunkyo-ku
Tokyo 113-8656, Japan

Leksono Probo Subanu, Ph.D.
Professor & Director
Center for Studies in Regional Development Planning
Gadjah Mada University
Jl.Socio Yustisia, Bulaksumur
Yogyakarta 55281, Indonesia

Johann Jessen, Dr.
Professor
Institute of Urban Planning and Design
University of Stuttgart
Keplerstr. 11, 70174 Stuttgart
Germany

Alain Motte, Doctorat d'Etat
Professor
Institute of Regional Planning
Université Paul Cézanne - Aix-Marseille III
2 Avenue Henri Poncet 13628 Aix en Provence Cedex I, France

Ethan Paul Seltzer, Ph.D.
Professor & Director
Toulan School of Urban Studies and Planning, Portland State University
PO Box 751, Portland
OR 97207-0751, USA

Cover photos:
(*Upper*) Yogyakarta, Indonesia; © Norihisa Shima
(*Lower*) Matsuyama, Japan; © Tetsuo Kidokoro

ISSN 1865-8504
ISBN 978-4-431-78146-2 ISBN 978-4-431-78147-9 (eBook)

Library of Congress Control Number: 2008923165

Springer is a part of Springer Science+Business Media
springer.com

Typesetting: Camera-ready by the editors and authors

Printed on acid-free paper

Preface

The Center for Sustainable Urban Regeneration (cSUR) at the University of Tokyo has undertaken research on sustainable urban regeneration and the governance of city regions. The objective of the research is to examine the characteristics of the emerging space of global city regions and to discuss possible strategies and forms of governance for those city regions, with a focus on the roles of regional cities as creative hubs of city regions.

Three International Workshops on Sustainable Urban Regeneration and Local Governance in Regional Cities were held in order to make a comparative study among city regions in Asia, Europe and the US. The first International Workshop was held in Yogyakarta, Indonesia on 4 March 2005 in collaboration with the Gadjah Mada University, and the second and third International Workshops were held in Tokyo with support from the Japan International Cooperation Agency (JICA) on 17 January 2006 and 12 January 2007 respectively. The commonalities as well as differences of the selected case city regions were discussed in depth and many useful insights and ideas emerged during the workshops. This book presents the results of these workshops. All contributors presented their papers in the workshops and expanded their initial analysis and ideas based on the discussions at the workshops and subsequent correspondences among the participants.

This book has combined empirical and comparative analysis on the governance of city regions in Asia, Europe and the United States, which, we believe, will add a new perspective to discussions on the sustainable future of cities and regions. We hope that the ideas conceived in this book will serve as a useful and thought-provoking resource for researchers, practitioners and students and those who are interested in the nature of emerging city regions and sustainable cities and regions.

Tetsuo Kidokoro

Acknowledgments

We would like to thank all of the contributors to this book as well as those who directly or indirectly participated in the comparative case studies for their contributions to the final results of the projects. We would like to give special thanks to Ann Tomoko Yamamoto for her dedicated work as an English editor for this book.

Contents

Preface *v*

Tetsuo Kidokoro

List of Contributors *xi*

Part I. Introduction

1. New Directions in Urban Regeneration and the Governance of City Regions..3

Tetsuo Kidokoro, Akito Murayama, Kensuke Katayama and Norihisa Shima

2. Strategy for a Sustainable Transportation System in City Regions: Strategic Approaches and Consensus Building.....23

Noboru Harata

Part II. China, Korea, Indonesia, Thailand

3. Governing Urban Development in Dualistic Societies: A Case Study of the Urban Region of Yogyakarta, Indonesia.........41

Leksono Probo Subanu

4. Problems and Prospects for Sustainable Regeneration of Traditional Markets in Yogyakarta City...........................61

Kawik Sugiana

5. De Facto Urban Regeneration: A Case Study of Chiang Mai City, Thailand...77

Niramon Kulsrisombat

6. Spatial Regeneration and Beyond: Daegu, Korea............101
Shi-Chul Lee and Seungkeun Park

7. Urban Form and Rail Transit Development in Dalian, China...127
Shengchuan Zhao and Zhiwei Yang

Part III. Japan

8. Urban Regeneration and the Shift of Planning Approaches: The Case of Japanese Regional Cities..........................145
Tetsuo Kidokoro

9. Activities of Local Organizations for Downtown Revitalization in Regional Cities...163
Fumihiko Seta

10. Sustainable Regeneration of a Car-dependent City: The Case of Toyama toward a Compact City...........................183
Kiyoshi Takami and Kiichiro Hatoyama

Part IV. Denmark, Germany, France

11. Copenhagen, Denmark: Urban Regeneration at Economic and Social Sustainability...203
Hans Thor Andersen

12. Regional Governance and Urban Regeneration: The Case of the Stuttgart Region, Germany.................................227
Johann Jessen

13. City-Region Planning and Urban Projects "Coexistence or Integration?": The Case of Marseille-Aix in France (1990-2006)...247
Alain Motte and Jérôme Dubois

Part V. America

14. **Regional Planning and Local Governance: The Portland Story...277**
Ethan Paul Seltzer

15. **Sustainable Urban Regeneration in Phoenix, Arizona: Implications for Multi-dimensional Governance............299**
Carlos Balsas

Part VI. Conclusion

16. **Towards Sustainable Regeneration of City Regions........323**
Tetsuo Kidokoro, Noboru Harata, Leksono Probo Subanu, Johann Jessen, Alain Motte and Ethan Paul Seltzer

List of Contributors

Carlos Balsas
Assistant Professor
School of Planning
Arizona State University
balsas@asu.edu

Jérôme Dubois
Professor
Institute of Regional Planning
Paul Cézanne University
j.dubois.iar@wanadoo.fr

Hans Thor Andersen
Lecturer
Institute for Geography and Geology
University of Copenhagen
Hta@geogr.ku.dk

Kiichiro Hatoyama
Visiting Professor
Department of Civil Engineering
The University of Tokyo
kii@civil.t.u-tokyo.ac.jp

Noboru Harata
Professor
Department of Urban Engineering
The University of Tokyo
nhara@ut.t.u-tokyo.ac.jp

Johann Jessen
Professor
Institute of Urban Planning and Design
University of Stuttgart
johann.jessen@si.uni-stuttgart.de

Kensuke Katayama
Assistant Professor
Department of Urban Engineering
The University of Tokyo
kenkata@urban.t.u-tokyo.ac.jp

Tetsuo Kidokoro
Associate Professor
Department of Urban Engineering
The University of Tokyo
kidokoro@urban.t.u-tokyo.ac.jp

Niramon Kulsrisombat
Lecturer
Department of Urban and Regional Planning
Faculty of Architecture
Chulalongkorn University
niramon.piadaeng@gmail.com

Shi-Chul Lee
Assistant Professor
Department of Public Administration
Kyungpook National University
shichul@knu.ac.kr

Alain Motte
Professor
Institute of Regional Planning
Université Paul Cézanne -
Aix-Marseille III
alain.motte@univ-cezanne.fr

Akito Murayama
Associate Professor
Department of Environmental Engineering and Architecture
Graduate School of Environmental Studies
Nagoya University
murayama@corot.nuac.nagoya-u.ac.jp

Seungkeun Park
Visiting Researcher
Research Center for Advanced Science and Technology
The University of Tokyo
skp104jp@ybb.ne.jp

Ethan Paul Seltzer
Professor & Director
Toulan School of Urban Studies and Planning
Portland State University
seltzere@pdx.edu

Fumihiko Seta
Associate Professor
Graduate School for Creative Cities
Osaka City University
seta@gscc.osaka-cu.ac.jp

Norihisa Shima
Assistant Professor
Department of Civil Engineering
The University of Tokyo
Tokyo, Japan
shima@civil.t.u-tokyo.ac.jp

Leksono Probo Subanu
Professor & Director
Center for Studies in Regional Development Planning
Gadjah Mada University
lsubanu@psppr-ugm.net

Kawik Sugiana
Professor & Director
Graduate Program in Urban and Regional Planning
Gadjah Mada University
putri_mpkd@yahoo.com

Kiyoshi Takami
Assistant Professor
Department of Urban Engineering
The University of Tokyo
takami@ut.t.u-tokyo.ac.jp

Zhiwei Yang
Research Assistant
Center for Transportation Research
Dalian University of Technology
duter@163.com

Shengchuan Zhao
Director
Center for Transportation Research
Dalian University of Technology
szhao@dlut.edu.cn

Part I

INTRODUCTION

1. New Directions in Urban Regeneration and the Governance of City Regions

Tetsuo Kidokoro, Akito Murayama, Kensuke Katayama and Norihisa Shima

1.1 Introduction

Cities should be recognized as social, economic as well as environmental systems formed through networks of cities, towns and villages rather than as isolated entities. Conceptualized as such, they are often called city regions. Central cities of city regions, which we call regional cities, play a decisive role in regional as well as national development. In an age of globalization, competition among such city regions is one of the major driving forces in their on-going transformation. Globalization may strengthen the sustainability of some city regions through enlarging their economic bases, or may weaken their sustainability through the loss of industrial competitiveness or loss of geopolitical importance. Indeed, regional cities are now at a crossroads of whether they decline or are regenerated under the impacts of globalization.

In many regional cities there are wide-ranging attempts to respond to the impacts of globalization and develop regional development strategies from the standpoint of sustainable development, though it is still unclear how such spatial strategies can work effectively. Our concern here lies in urban regeneration and strategic spatial planning at the city-region level. The reason is that, we believe, the sustainable regeneration of cities as creative regional centers will play a decisive role in the sustainable development of city regions as a whole, but only with viable regional spatial strategies that strengthen the network of cities and their hinterlands. This chapter overviews arguments on the formation of sustainable city regions. The experiences of EU countries are highlighted, since city regions have been extensively discussed in these countries. Trends of discussions in Asia as well as the United States are overviewed as well.

1.2 Space for Strategic Spatial Planning

1.2.1 From 'Region' to 'City Region'

Recently the cognitive framework of regional development in terms of spatial dimensions has undergone a critical change throughout the world, particularly in European countries. In Europe, discussions on spatial planning since the early 1990s have responded to the need to rescale space due to the formation of the EU as well as impacts of globalization. During the 1980s, particular emphasis was placed on the 'region' as the historical, cultural and economic entity at the sub-national scale. Since the 1990s, however, the concept of 'city region' has come to the fore (see Fig. 1-1). In the EU regional policies *Europe 2000* (1991), *Europe 2000+* (1994) and Urban Agenda in Europe (1997), the policy focus in regional development has shifted and cities have emerged as vital centers of regional development from the perspective of a competitive and innovative Europe (Herrschel 2002). Factors behind the emergence of the city region concept include increasing competition among cities and regions due to the EU's single market and single currency, as well as the need to create a more knowledge-based industrial structure in response to globalization. Thus,

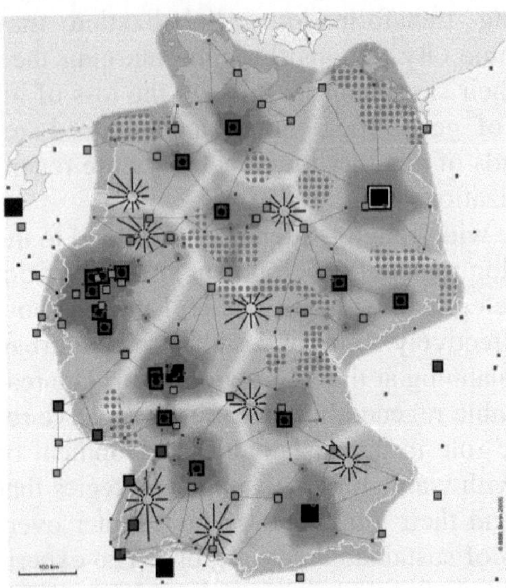

Fig. 1-1. National Spatial Development Guideline, Germany, based on the concept of city region (2006) (Source: Ministerial Conference for Regional Planning (MKRO) (2006))

since the 1990s, there has been growing consensus that the growth of economically vital cities leads regional, national and Europe-wide economic development.

The European Spatial Development Perspective (ESDP), confirmed in an informal EU ministerial meeting in 1999, embodies the recent focus on the city region concept. Importantly, the spatial concept developed in ESDP has influenced EU countries' spatial policies (Counsell et al. (2003)). ESDP's main objectives are: economic and social cohesion; conservation of natural resources and cultural heritage; and more balanced competitiveness within the European territory. With these objectives, ESDP advocates three key policy guidelines: the development of a balanced and polycentric urban system and a new urban-rural relationship; securing parity of access to infrastructure and knowledge; and sustainable development, prudent management and protection of nature and cultural heritage.

National-level regional development policy in many countries has also shifted from supporting economically weak areas prior to the 1970s to emphasizing the role of successful cites in strengthening nationwide innovative capacity since the late 1980s (Salet et al. (eds) (2003)). Under competitiveness-oriented economic development policies, many countries are pursuing new regional development strategies to enhance international accessibility through enlarging international airports, providing high-speed access such as express rail and motorways, as well as improving 'knowledge infrastructure' such as universities, research institutes and cultural facilities to strengthen competitiveness and attract innovative firms. This new thinking about regional development emphasizes the role of cities as hubs of international knowledge networks and increasingly recognizes the need for urban regeneration.

1.2.2 America 2050 Movement

In the United States, national planning is widely believed to be an un-American activity. There is, however, a national initiative called "America 2050" to meet the nation's infrastructure, economic development and environmental challenges as it prepares to accommodate 120 million additional people by the year 2050. The Regional Plan Association, an independent, not-for-profit regional planning organization for the 31-county New York-New Jersey-Connecticut region, convened the National Committee for America 2050, a coalition of regional planners, scholars, and policy-makers to develop a framework for the nation's future growth. The development of a national framework is recognized as urgently needed in a time of rapid technological, geopolitical, demographic, and environmental change. As never before, the nation's economic prospects are being threatened by global competitors, all of whom have created long-range strategies for growth and competitiveness (America 2050); (Regional Plan Association (2007)).

The emergence of mega-regions is considered as a key to the country's growth since the nation's most rapid population growth, and an even larger share of its economic expansion, is expected to occur in 10 or more emerging mega-regions: large networks of metropolitan regions, each mega-region covering thousands of square miles and located in every part of the country. Major categories that define mega-regions are: environmental systems and topography, infrastructure systems, economic linkages, settlement patterns and land use, and shared culture and history (Regional Plan Association (2006)).

The recognition of emerging mega-regions enables cooperation across jurisdictional borders to address specific challenges experienced at this scale. One way mega-regions can prepare for future population pressures is securing resources to make investments in high-speed rail and other mobility infrastructure. Other ways include: protecting environmental resources, coordinating economic development strategies, and making land use decisions that integrate all of the above (Regional Plan Association (2006)).

1.2.3 Formation of Global City Regions

Along with the arguments on city regions, attention has shifted to development corridors/clusters which are formed through the conurbation of city regions across national boundaries. They are often referred to as "global city regions" (Scott (2001)). Knowledge has the characteristic that it increases as it is consumed, and so the knowledge-based industry typically generates increasing-returns effects. In an age of globalization, the mobility of people, goods and capital has increased dramatically and thus the competition for location of industry has also increased. On the other hand, the competitiveness of firms in the global market deeply relies on regional settings such as institutional set-up, training and education capacity, research and development functions, as well as time-proximity to international customers. In addition, language barriers restrict the international mobility of knowledge workers to some extent; thus, their mobility tends to be largely influenced by the national hierarchy of cities, rather than the international network of cities. Face-to-face exchange of information and personal networks are important in knowledge-based industries. At the same time, international knowledge networks play decisive roles in generating innovation, thereby further strengthening the role of international hub cities (Simmie (ed) (2002)).

As the space generated in these conflicting conditions, the global city region is characterized by conversion and diversion at the global scale (Scott (2001)). As shown in Fig. 1-2, development corridors/clusters beyond national borders are observed in many parts of the world. Indeed, internal trade within each supranational region, particularly among East Asia countries, is steadily increasing in response to the formation of such global city regions.

1.2.4 New Theories of Agglomeration Economies

Though the EU's regional development policy emphasizes balanced development from social, economic and environmental perspectives, many scholars argue that the main focus of regional development in each country lies in economic policy, in particular, the enhancement of regional competitiveness and innovation capacity (Herrshel and Newmann (2002)). Against this background, since the 1990s many arguments have proposed new theories on agglomeration economies, such as industrial district theory, innovative milieu theory, transaction costs theory, and industrial cluster theory. According to these new theories, formal and informal networks in the region are essential as the basis of innovation. They are represented by such factors as: customer-client relations, education and training, exchange of knowledge, competition and cooperation among SMEs, the local culture of cooperative working, etc. Reflecting such insights on agglomeration economies, the focus of regional economic development strategies has recently shifted to innovation-centric and place-marketing strategies, including: the support of agglomeration-specific R&D, the provision of high-risk capital to venture firms, the preservation of specific infant industry, investment in education and training of workers, the cultivation of collaborative inter-firm relations, and the development of markets for local products (Scott (2001)).

From the viewpoint of urban regeneration, particular attention has been paid to R. Florida's creative economy theory. Florida (2002) argues that the source of regional development lies in 3 T's (technology, talent and tolerance), based on his analysis of creative centers in the United States. According to his arguments, regional economic development is in part generated through the attraction of "creative class" people composed of scientists, engineers, artists, musicians, designers and knowledge-based professionals. Thus, it is important to invest in creating vibrant, attractive cities offering "interrelated sets of experiences" like dynamic and

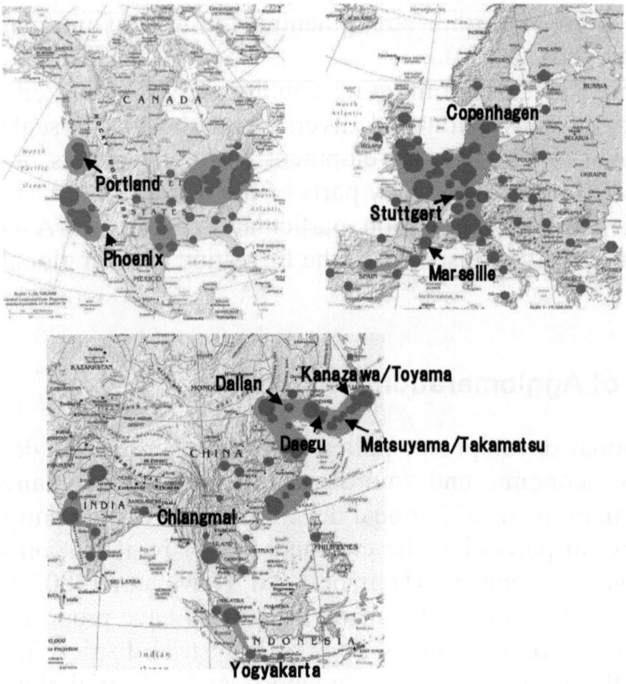

Fig. 1-2. Formation of global city regions

participatory street-level music scenes open to everyone, which match the life style of creative class people. In contrast, most US cities invest in sports stadiums, freeways, urban malls and theme parks like tourism and entertainment districts, which are not attractive to creative class people.

1.3 Governance for City Regions

1.3.1 Formation of Regional Governance

Governance is seen as "a flexible pattern of public decision-making based on loose networks of individuals" (John (2001)). It is referred to in the context of the shift from government-style decision-making systems to governance-style decision-making systems. There are clear signs of this shift in many countries. In European countries, since the 1980s various attempts have been made through collaboration between government agencies and private sector associations. Some argue that these collaborations tend to focus on strengthening competitiveness, and the public decision-making process becomes diversified and somewhat blurred due to the establishment of wide-ranging organizations such as ad hoc bodies and appointed organizations often called quangos (quasi-autonomous non-governmental organizations) (Salet et al. (eds) (2003)).

In England, metropolitan councils such as the Greater London Council (GLC) were abolished under the Thatcher government. In order to fill the vacuum of metropolitan governance, since the 1980s an informal arena of public decision-making was development through a variety of public-private partnerships (PPPs) such as London First and the London Pride Partnership. The institutionalization of regional governance began in the 1990s, partly in response to EU regional policy and funding (Thomas et al. 2000). Nine regions were newly created as administrative areas with formal regional institutions. The Government Office of the Region (GOR) integrated sector-based national government offices, the Regional Development Agency (RDA) serves as a PPP-style agency responsible for the formulation and implementation of the Regional Economic Development Strategy, and the Regional Assembly/Chamber is composed of local authority representatives, private sector associations, and voluntary associations from environmental and social sectors. Among them, the Greater London Authority (GLA), established in 2000, has a unique character. GLA has a directly elected mayor and also a directly elected assembly, unlike other regions. Thus, GLA enjoys wider power and a stronger role than institutions in other regions. For example, applications for development strategically important to London must be passed to the mayor by boroughs that receive the applications. The mayor can give a direct refusal if the development project does not conform to GLA strategies.

In France, the decentralization law of 1982 (*Loi Defferre*) established regions (*région*) as local governments and in 1986 the first regional elections for regional councils were conducted. The region has a crucial role in coordinating investments of municipalities (*commune*), in particular, through the State-Region Planning Agreements. In the case of the *Ile-de-France* region (the national capital region), the first State-Region Planning Agreement was the state's initiative. However, the *Ile-de-France* (CRIF) regional council began formulating its own regional development plan (SRADT) and became able to address its own strategy in negotiations with the state, thereby strengthening the CRIF's role as coordinator of the region's municipalities (Lefèvre (2003)). The PPP-style Regional Development Agency (ARD) was created in the Ile-de-France region in 2001, an important development from the viewpoint of regional governance.

Germany has a hierarchical spatial planning system under its federal system. The regional planning associations composed of municipalities are responsible for formulating and enforcing regional plans covering city-region areas, though regional planning systems vary from state to state. The state government supervises regional planning associations; thus, it can be said that spatial planning is well established within the government systems in Germany. Yet, there are also various attempts to shift from government to governance. The Stuttgart city region is a remarkable case. Stuttgart suffered economic recession in the early 1990s and experienced extensive job loss. During that period, regional consensus emerged on the importance of developing regional competitiveness strategy. In 1994, the Stuttgart Regional Association (*Verband Region Stuttgart*: VRS) was established, which has a directly elected regional council. VRS voted for an innovative concept to combine spatial planning with economic development. To realize this concept, VRS established the Economic Development Agency for Stuttgart Region (WRS) with the participation of local municipalities, chambers of commerce, trade unions and various associations to implement competitiveness and innovation strategies (Heeg (2003)). That same style of governance formation at the city region scale is seen elsewhere in Germany.

1.3.2 Regional Corporatism

Many scholars point to the formation of regional corporatism as the major characteristic of regional governance in Europe. Corporatism refers to the political situation where public decision-making is made through negotiation among a few corporate bodies. The city-region scale is much larger than the municipal scale, and so powerful public and private key players who can mobilize resources at the regional scale have more opportunities to influence public decision-making. Thus, regional relationships tend to be dominated by economic considerations. Citizens and less well organized environmental or business groups will find few opportunities (Vigar et al. (eds) (2000)). In England, where the region as an administrative unit is a new creation, regional actors tend to be limited to local authorities, powerful business associations, and a few national-scale environmental associations (Marshall (2004)).

This is particularly the case in large-scale urban regeneration projects. Such projects are conducted with funding from the EU, national governments, and local governments, in many cases, on public land (Couchi et al. (2003)). Following the trend of deregulation, local authorities often deviate from planning regulations as the result of negotiations with key players such as developers, business associations, and national governments in order to attract private investment. In situations like these, the decision-making process tends to become obscure.

1.4 Urban Regeneration and Spatial Planning

1.4.1 Project-based Approach for Urban Regeneration

Faced with large-scale job loss under the impacts of globalization, city authorities have increasingly sought to develop strategies to attract private investment through private-public sector collaboration. This developmental model embodies the departure from the post-war welfare state model of regulation and government provision, and is often referred to as the shift from 'managerial city' to 'entrepreneurial city.' As for planning style, a project-based approach is typically used in the entrepreneurial city model. In this approach, the role of plans is rendered less significant and the role of projects is emphasized, thereby restructuring urban space via 'planning-through-projects' rather than 'planning-through-plans' (Healey (1997)). Spatial plans become a sort of the shopping lists of projects or merely a stapled aggregation of policies of different sectors. Projects tend to be prestigious/symbolic projects such as waterfront regeneration, opera houses, international airports, high-speed rail stations, international convention/messe complex, or international sports/culture events, etc. It is expected that, through the execution of projects, the city's image is enhanced and successive investment is attracted to the city, thereby regenerating the city.

Examples of top-down national projects include millennium projects in London, the Grand *Projet* in Paris, and various capital relocation projects in Berlin. The IBA *Emscher* Park (1989-1999) in Germany's Ruhr region is a remarkable bottom-up project, where more than 100 symbolic cultural and ecological projects have been developed through collaboration between public and private agencies in order to symbolize the transition of the coal and iron region into an innovative economic and cultural center.

1.4.2 Spatial Strategic Planning Approach

While project-based approaches have become a mainstream planning style, since the 1990s there has also been a strong trend of strategic spatial planning, which emphasize plan-led approaches (Healey et al. (eds) (1997)).

Limitations to the project-based approach underpin this revival: uncertainty in property markets, complaints from landowners and developers on unfair treatment, increasing oppositions from environmental and community groups, unclear decision-making through closed-door negotiation processes, and growing demand from citizens for more involvement (Healey (1997)). Yet, of course, the newly emerging spatial planning cannot be the same as before. Indeed, the focus of spatial planning has shifted from regulation and allocation of space for anticipated growth to stakeholder collaboration and multi-sector strategy formulation in order to stimulate urban development. This shift of planning style can be described as the shift from 'plan-command-control' to 'plan-monitor-manage' planning style.

In line with this trend, institutional reforms on spatial planning are currently underway in many European countries. In England, the Planning and Compulsory Purchase Act of 2004 comprehensively revised the framework of planning systems into two-tier plans at the regional level and the local levels: the Regional Spatial Strategy (RSS) to be formulated by Regional Planning Bodies (Regional Assembly or the mayor in the case of Greater London) and the Local Development Framework (LDF) formulated by local authorities (Districts). The RSS is to move beyond traditional land use regulatory planning to comprehensively incorporate multiple sectors such as housing, regional transport, regional economic development, environmental strategy and social dimensions. Furthermore, the RSS is stipulated as the long-term framework for the Regional Economic Development Strategy formulated by RDA. There is a clear intention to knit different dimensions of sustainable development together through spatial planning (Marshall (2004)). LDF is also stipulated to have strong links with community strategy formulated by local authorities, going beyond traditional land use regulations to serve as a spatial framework for sustainable development (Counsell et al. 2003).

In France, where more than 30,000 municipalities (*commune*) exist and there is a strong tradition of commune autonomy, various efforts to institutionalize inter-communal cooperation have largely failed, except for a few metropolitan areas such as Lyon and Lille which have a tradition of cooperation among communes. However, the Inter-communal Law, 1999 (*Loi Chevènement*) and the Urban Solidarity and Regeneration Law, 2000 (*Loi SRU*) provide new prospects for the formulation of strategic spatial planning at the metropolitan level (Booth (2003)). The Inter-communal Law established a new framework of metropolitan communal associations through the incentives of common taxation of business tax and a lump sum subsidy at the time of corporation. The *Loi SRU* shifts to a multi-sector strategic spatial planning approach, as seen in its three components: reshaping the framework of spatial coordination of urban developments at the metropolitan level; integration of land use planning and provision of social housing; and integration of land use planning and urban transport planning.

As for spatial coordination at the metropolitan level, the *Loi SRU* replaced the 1967 two-tier planning systems (SD at the metropolitan level and POS at the local level) with new two-tier planning systems (SCOT and PLU respectively). The roles and functions of SCOT are substantially stronger than the former SD. In the absence of a SCOT, no development is allowed on unbuilt land, whether zoned for future development or not in PLU, in areas up to 15 km from the commune with a population over 15,000. Another important feature is commercial buildings with floor area exceeding 5,000 m^2 are regulated directly by SCOT. Most importantly, it was stipulated that the designation of ZAC (designated action area; planning tool for communes to conduct large-scale urban development projects) should be in accordance with SCOT. Formally ZAC could be independently designated by communes.

Inter-communal cooperation in spatial planning has also been strengthened in the SCOT system. In the case of the former SD, it was not obligatory for metropolitan areas to create an SD, and inter-communal associations created to formulate the SD were to be dismantled after they finished making the plans. In the case of SCOT, all the inter-communal associations created under the Inter-communal Law, 1999 are obligated to formulate SCOT. Even when the new inter-communal associations are created especially for SCOT, they are maintained for monitoring and revision after finishing plan-making, reflecting the new strategic planning concept of 'plan-monitor-review.' However, there are still significant limitations. The spatial setting of 15 km from towns with the populations of 15,000 or more may lead to the fragmentation of city regions into several inter-communal cooperation areas.

1.4.3 Smart Growth

In United States, in line with the spatial strategic planning approach, smart growth is becoming mainstream in many of states. According to the U.S. Environmental Protection Agency, the smart growth concept covers a range of development and conservation strategies that help protect the nation's natural environment and make its communities more attractive, economically stronger, and more socially diverse (U.S. Environmental Protection Agency). The birth of smart growth is difficult to pinpoint, but its roots include not only the growth management revolution of the 1970s and the 1980s in the coastal states (both east and west), but also the new urbanism movement of the 1990s led by architects and the worldwide attention to sustainable development after the publication of "Our Common Future" in 1987 by the World Commission on Environment and Development (the Brundland Commission). The smart growth movement is a national movement led by a network of environmental groups, historic preservation organizations, professional organizations, developers, real estate interests as well as local and state government entities. In 1996, the U.S. Environmental Protection Agency established the Smart Growth Network (SGN) with the above partners to: raise public awareness of how smart growth can improve quality of life; promote smart growth best practices; develop and share information, innovative policies, tools and ideas; and cultivate strategies to address barriers to and advance opportunities for smart growth. Through a cooperative agreement, SGN is managed by the International City/County Management Association and the U.S. Environmental Protection Agency (Smart Growth Network).

In communities across the United States, there is a growing concern that current development patterns, dominated by so-called sprawl, no longer reflect long-term visions regarding cities, existing suburbs, small towns, rural communities, and wilderness areas. Though they continue to support growth in terms of population and economy, communities are questioning the costs of abandoning infrastructure in the city, only to rebuild it further out. In general, smart growth invests time, attention, and resources in restoring community and vitality to city centers and older suburbs. New smart growth is more town-centered, transit and pedestrian oriented, and has a greater mix of housing, commercial and retail uses. It also preserves open space and many other environmental amenities. The ten principles adopted by SGN clearly present the vision of smart growth:

- Create range of housing opportunities and choices
- Create walkable neighborhoods
- Encourage community and stakeholder collaboration
- Foster distinctive, attractive communities with a strong sense of place
- Make development decisions predictable, fair and cost effective
- Mixed land uses
- Preserve open space, farmland, natural beauty and critical environmental areas
- Provide a variety of transportation choices
- Strengthen and direct development towards existing communities
- Take advantage of compact building design

The handbooks "Getting to Smart Growth" volume one and two published by SGN further describe concrete techniques for putting the above principles into practice. The policies and guidelines presented in these primers have proven successful across the United States, ranging from formal legislative or regulatory efforts to informal approaches, plans, and programs. The handbooks serve as a roadmap for states and communities that have recognized the need for smart growth but are unclear on how to achieve it (Smart Growth Network and International City/County Management Association (2002)); (Smart Growth Network and International City/County Management Association (2003)).

In many respects, these principles and implementation measures seem not so different from those of growth management, new urbanism or sustainable development, or just good planning. Conventional planning tools such as comprehensive planning, subdivision control, zoning, transfer of development rights, incentive bonus, and design review are still widely used in smart growth initiatives. However, as Knaap (2005) explains, based on the case of Maryland, the smart growth movement contrasts with its antecedents in the following ways: incentives for implementation; integrated transportation and land use policy; insurgency and advocacy; innovative policy instruments; and institutional reform.

1.4.4 Two Shifts: Promotion of Participation and Decentralization in Asian Countries

Since the 1980s, important shifts have occurred in Asian countries, namely the promotion of participation and decentralization. Both trends have important implications for local governance.

Former developmental states like Indonesia, Korea, the Philippines, and Thailand became democratic states in the 1980s, in step with the emergence of a new middle class. Korea typifies this change. The Asian economic crisis, starting in 1997 in Thailand, accelerated the trend. In Indonesia, Soeharto's regime collapsed in 1998, resulting in the promotion of citizen participation in decision making processes in various fields (Institute of Social Science (ed) (1998)). There is increasing numbers of NGOs collaboratively working with governments in many Asian countries.

Decentralization is also an important shift in Southeast Asian countries. Though the degree varies from country to country, especially since the 1990s, decentralization is being promoted, even in countries with transitional economies like China and Vietnam. For example, in Indonesia, 'big-bang' decentralization was promoted in 2001, and local government has much more authority than before (World Bank (2005)).

These two shifts can be an opportunity for regional city development policy through the regeneration of local governance. However, many countries are still preparing the institutional framework for encouraging participation and decentralization, and the process is trial and error for both government and citizens. Some attempts at the local level succeed, and others do not. Various lessons need to be drawn from development experiences.

1.5 Urban Regeneration in Regional Cities in Japan

In Japan, discussions on urban regeneration in regional cities focus on city center revitalization. It is widely accepted that policies promoting shopping districts in the city center are not enough to solve the urban regeneration issue. However, an effective framework for regional planning does not exist in Japan. Thus, spatial framework and cultural creativity are being discussed.

1.5.1 Compact City Concept

The compact city concept is currently mainstream urban policy in Japanese cities. This concept is reflected in amendments to the city planning law and downtown revitalization law, which aim to update the land use system to regulate urban sprawl, put city functions together into urbanized areas, and rebuild the central government's support system for city center revitalization. In principle, large retail/attracting facilities (floor area over 10,000m^2) can be located only in commercial districts, neighborhood commercial districts and quasi-industrial districts in urbanization promotion areas. At the same time, municipalities are required to create corresponding city planning policies regulating suburban development when they want their downtown revitalization plans to be approved by central government. This means that policies to create compact cities and revitalize city centers are promoted in the integrated way.

Examples include the urban policies in Aomori City (Aomori Prefecture), which aim to form a mono-centric compact urban structure led by redevelopment projects in the city center and zoning into inner, mid and outer areas, and policies in Toyama City (Toyama Prefecture), which aim for a "skewer and dumpling" urban structure led by transit-oriented development. There are also other local governments which have their own ordinances on land-use regulation for growth management, such as Fukushima Prefecture, which enacted an ordinance to regulate the location of large retail facilities.

1.5.2 Culture and Creativity

Given the change from a post-industrial to a knowledge-based society, utilization of art and culture, intellectual stimulation of creative peoples, and creating attractive living conditions are increasingly seen as important keys to urban regeneration. This trend reflects the transition of urban policies in European countries. Sasaki (2003) defines the creative city as a place "which has creativity in culture and industry, innovative and flexible urban economy system, and acts as a 'place of creativity' that contributes to creatively solving global environmental issues or local community issues." He cites Kanazawa City as a typical example in Japan.

1.5.3 Regional Coordination Measures in Japan

The 2nd comprehensive National Development Plan (1969) established cooperative measures among municipalities (known as the "wider living area" concept) such as the Wider Administrative Area for Local Townships and the Local Life Area for Daily Life. However, these did not function simply due to the lack of regional planning bodies. In 1995, the administrative union system among neighboring municipalities was introduced and was expected to function as regional planning bodies for spatial planning as well. This system has not been utilized. Instead, the merger of municipalities has been strongly promoted by the central government. About 3,000 municipalities have been merged into about 1,800 as of 2007. If the new boundaries of the expanded municipalities reflect the actual metropolitan areas, this will provide an effective framework for regional planning. Unfortunately, this is not the case, since the merger of municipalities is decided politically. Indeed, there are no guidelines on the mergers in terms of spatial considerations.

Aside from the merger of municipalities, the new provision in the amended city planning law of 2006 is an important step toward regional coordination. Consultation with and consent from the prefecture is needed when large-scale shopping centers are located in suburban areas, and prefectures can consider opinions not only from the concerned municipality but also neighboring municipalities. This provision might strengthen the role of prefectures in regional coordination among municipalities.

1.5.4 *Machizukuri* Movement

In Japan, volunteer activities and public participation have been promoted from the 1980s. Activities have appeared that not only oppose government proposals, but also propose alternative ideas and build partnerships. Those activities are called *machizukuri* (town-making). Particularly, the role of the voluntary sector has been recognized since the Great Hanshin-Awaji Earthquake in 1995, when large numbers of volunteers worked during the immediate rehabilitation period. This led to the establishment of the Law to Promote Specified Nonprofit Activities in 1998. In the field of urban regeneration, the partnerships and networks among diverse stakeholders are also emphasized.

1.6 Issues to be Discussed

Space of City Region

What spatial characteristics does the city region have? According to new theories of agglomeration economies, city regions may be regarded as spaces of intricate knowledge networks in spatial proximity characterized by face-to-face interrelations. Cities should be regarded as creative centers in the knowledge-based economy. Following this idea, the city region can be seen as spaces within about 50km from the city center, where face-to-face knowledge networks can be realized. However, does this spatial arrangement fit with other dimensions of sustainable development, namely, environmental and social aspects? Moreover, is it suitable for the involvement of a variety of stakeholders and citizens?

Urban Regeneration and Strategic Spatial Strategy

Faced with the need to change the industrial structure under globalization, many large-scale urban regeneration projects have been conducted in areas like waterfronts as symbolic projects to enhance the image of the city and attract investment. These urban regeneration projects are typically based on the project-based approach. Yet, as seen in the revival of strategic spatial planning in Europe and the smart growth movement in the United States, the limitations of the project-based approach from the viewpoints of environmental sustainability, social dimensions, and economic uncertainty are increasingly becoming apparent. A question arises: how can urban regeneration projects be positioned in strategic spatial planning at the city-region scale? And, how can spatial planning fulfill the expected missions of coordinating multi-sector policies and mediating inter-municipal spatial conflicts? In terms of spatial composition, monocentric city regions seem to have an advantage over polycentric cities in terms of consensus building. Generally speaking, societies like Europe which have a tradition for government-imposed limits on individual property rights based on collective interest will be at an advantage compared with other parts of the world. Government policies will be also vary according to the kinds of structural change confronting the city, that is, whether the economy is in a developing stage, or in a stage of structural industrial change like in developed countries. The role of spatial planning should naturally change as government policies reflect the development stage of the city. But how?

Local and Regional Governance

The final question raised here concerns the styles of local and regional governance. Ideally, the creation of inter-municipal associations whose councils are directly elected is most desirable. However, various styles of institutional set-up reflect the social and political situations in respective countries. Regardless, establishing informal collaborative networks among national government agencies, local governments, private sector associations and other environmental and social groups is the first step for the formation of local and regional governance. In particular, at the regional level, there is a danger that a few limited groups such as business associations or national government agencies who can mobilize resources at the regional level will unduly influence public decision-making. On the other hand, the involvement of citizens may end in just a formality because citizens may not be familiar with regional issues. In this regard, it is important to relate the local governance at the municipality level with regional governance at the city-region level in order to smoothly create arenas for citizen participation and collaboration among stakeholders.

References

Booth, P. (2003) “Promoting radical change: The Loi Relative à la Solidarité et au Renouvellement Urbains in France”, *European Planning Studies*, 11(8)

Couchi, C., Fraser, C. and Percy, S. (2003) *Urban Regeneration in Europe*. Blackwell

Counsell, D., Haughton, G., Allmendinger, P. and Vigar, G. (2003) “New directions in UK strategic planning”, *Town & Country Planning*, January

Desfor, G. and Jorgensen, J. (2004) “Flexible urban governance; the case of Copenhagen’s recent waterfront development”, *European Planning Studies*, 12(4)

Fishman, R. (2007) “1808-1908-2008: National Planning for America”, *America 2050* (Regional Plan Association), July 8-13, 2007

Florida, R. (2002) *The Rise of the Creative Class*. Basic Books

Healey, P. (1997) *Collaborative Planning: Shaping places in fragmented societies*. University British Columbia Press
Healey, P., Khakee, A., Motte, A. and Needham, B. (eds) (2003) *Making Strategic Spatial Plans: Innovation in Europe*. Routledge
Heeg, S. (2003) "Governance in the Stuttgart metropolitan region", in Salet, W., Thornley, A. and Kreukels, A. (eds) (2003) *Metropolitan Governance and Spatial Planning: Comparative Case Studies of European City-Regions*. Spon Press
Herrschel, T. and Newman, P. (2002) *Governance of Europe's City Regions: Planning, Policy and Politic*. Routledge
Institute of Social Science, The University of Tokyo ed. (1998) *Developmentalism*. University of Tokyo Press (in Japanese)
Iwasaki, N. et al. (eds) (2006) *The Policies and Governance in Regional Society*. Toshindo Publishing (in Japanese)
John, P. (2001) *Local Governance in Western Europe*. SAGE Publications
Knaap, G. J. (2005) "A Requiem for Smart Growth?", in Daniel Mandelker (ed) *Planning Reform in the New Century*. APA Planners Press
Kumssa, A. and McGee, T.G. (eds) (2001) *New Regional Development Paradigm Vol. 1: Globalization and the New Regional Development*. Greenwood Press
Lefèvre, C. (2003) "Paris: Ile-de-France region" in Salet, et al. (eds) (2003) *op. cit*
Marshall, T. (2004) "Regional planning in England", *Town Planning Review*, 75(4)
Regional Plan Association (2006) *America 2050: A Prospectus*
Salet, W., Thornley, A. and Kreukels, A. (eds) (2003) *Metropolitan Governance and Spatial Planning: Comparative Case Studies of European City-Regions*. Spon Press
Sasaki, M. (2003) *Challenges toward Creative City*, http://www.creativecity.jp/index.html (in Japanese)
Scott, A.J. (2001) "Globalization and the Rise of City-regions", *European Planning Studies*, 9(7)
Simmie, J. (ed) (2002) *Innovative Cities*. Spon Press
Smart Growth Network and International City/County Management Association (2002) *Getting to Smart Growth: 100 Policies for Implementation*
Smart Growth Network and International City/County Management Association (2003) *Getting to Smart Growth: 100 More Policies for Implementation*
Stramback, S. (2002) "Change in the innovation process: New knowledge production and competitive cities-the case of Stuttgart", *European Planning Studies*, 10 (2)
Thomas, K. and Roberts, P. (2000) "Metropolitan strategic planning in England", *Town Planning Review*, 71 (1)
Vigar, G., Healey, P., Hull, A. and Davoudi, S. (2000) *Planning, Governance and Spatial Strategy in Britain: An Institutional Analysis*. Palgrave Macmillan
World Bank (2005) *East Asia Decentralizes: Making Local Government Work*. The World Bank

U.S. Environmental Protection Agency, http://www.epa.gov/smartgrowth/ (Accessed in 2007)
Smart Growth Network, http://www.smartgrowth.org/ (Accessed in 2007)
America 2050, http://www.america2050.org/ (Accessed in 2007)
Regional Plan Association, http://www.rpa.org/ (Accessed in 2007)

2. Strategy for a Sustainable Transportation System in City Regions: Strategic Approaches and Consensus Building

Noboru Harata

2.1 Introduction

In this chapter we explain approaches to developing a strategy for a transportation system that will provide sustainable mobility. The chapter explains the strategic planning process from goal setting to plan making and implementation, and then examines the relationship between transport strategy planning and decision-making processes. Then, we will discuss setting sustainable goals for transportation in city regions, and explain a strategy for rearranging the spatial development pattern. Finally, we cover possible outcomes of the strategic planning process, such as suggestions for institutional reform, recognition of a need for consensus building regarding the strategic vision developed, and the creation of bodies for local governance in city regions.

2.2 Basic Framework for Developing Transportation Strategy

2.2.1 The Strategic Planning Process

One important current issue in transport planning is the construction of a realizable and effective transportation strategy that will result in a system of sustainable mobility. Achieving this requires attention to the basic processes of transportation planning. The process of drafting and implementing a transportation strategy can be broken down into the following steps:

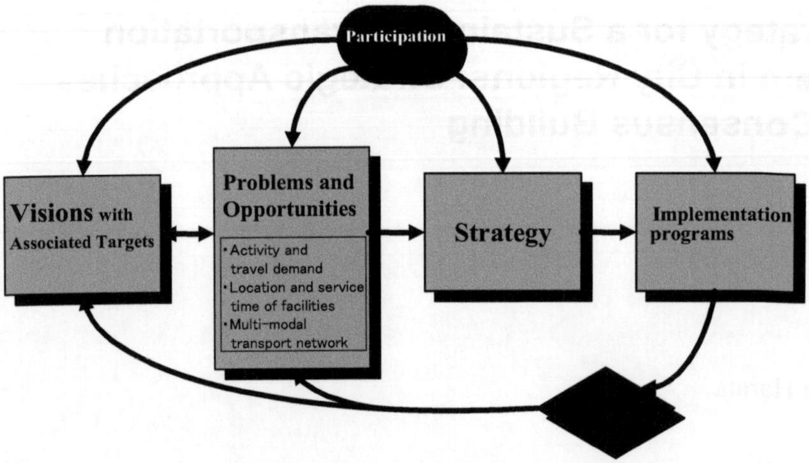

Fig. 2-1. Planning process of transport strategy

(1) establishing visions and targets, (2) understanding the present circumstances and funding opportunities of improvement, (3) examining and comparing alternative transportation strategies, and (4) implementing the best strategy for the particular circumstances. This process for formulating a strategy corresponding to policy targets has often been reviewed in infrastructure planning, including roads and railways. It is also highlighted in the Strategic Choice Approach, a concept that is relatively well known. However, strategic decisions corresponding to policy targets are seldom actually reached through this process, indicating the need for a clearer process that can be commonly applied.

One of the characteristics of this process is that it attaches importance to the complete process involving not only plan drafting but also plan implementation. This can be understood as a reflection of a shift to focusing on the stream of time, which requires a reorientation from considering a 'plan' (static object) to looking at 'planning' (dynamic activity). Another characteristic is the idea of 'management' as used in business, where a policy goal cannot be achieved until the planning process of plan, implement, and review is completed.

This is consistent with the way that Plan-Do-See (PDS) and Plan-Do-Check-Action (PDCA) cycles are currently applied in the planning field, as in the business field. Here, ongoing review is regarded as very important.

2.2.2 The Decision-making Process

In this section we consider what information is necessary to support decision making in the contemporary transportation planning context, recognizing that the provision of this necessary information is an important role for thc responsible transportation planner. The responsible transportation planner needs to present planning information in a straightforward manner that can be understood by the heads of local governments, the legislature and citizens.

Ohta has pointed out that the planning process is easier to understand when it is divided into two major aspects: the first is the decision-making process, and the second is the plan-making process to draft, evaluate and implement alternatives to be proposed and reviewed in the course of the decision-making process. The planning process itself is divided into five steps: (1) confirmation of the problem, (2) the drafting of a plan, (3) a review of the policy and draft and the finalization of the plan, (4) the implementation of the plan, and (5) evaluation and feedback. The decision-making process and plan-making process must be constantly interwoven at each step.

He has also pointed out the role of the responsible transportation planner:

> *The role of the responsible transportation planner is to provide the information necessary for decision-making during each step of the decision-making process. The information provided is not limited to the drafting of a specific alternative. It is important to support decision-making by providing timely and accurate information that includes an analysis of the present situation, the problems with the current traffic system and service, the progress and an evaluation of the existing policy and business, and problems and opportunities expected in the future, all things to meet the needs of the decision-maker (Ohta (1988))*

On the other hand, Muto, who has explained the policy process (Muto (2003)), has proposed dividing the policy process more minutely, into eight steps, as follows: (1) discovering the problem, (2) the issue of choices among public/private sectors, (3) pursuing problem solving techniques, (4) adjustments required in organization (5) the social process of implementing the decision, i.e., consensus building, (6) execution, (7) evaluation, and (8) feedback. The basic flow is the same as in the five step process. Items (2), (4), and (5) have been added. The relationship of these processes is considered below.

Item (2), the issue of choices among public/private sectors, can be seen as an important element in the planning framework, and it is included in step (2), the drafting of the plan, in the five-step process. Specifically, this item points to the problem of determining which sectors --- private corporations, nonprofit agencies, or government --- are to have which impact. This issue was not at the forefront when long-range transport strategies were first constructed, since the government sector played a key role in establishing the plan framework. However, nowadays creating a transport strategy for the future requires a proper combination of solutions involving market mechanisms driven by the corporate sector, solutions involving self-help efforts by nonprofit entities, and involvement of the government sector which contributes tax money. Thus, it has become important to consider implementation by various sectors during the review of a proposed transportation strategy.

In addition, step (4), adjustments required in organization, is added to clarify in the planning draft the legal grounds of the project, the sources of revenue, and the personnel involved. This makes it possible to consider the feasibility of the following issues that arise during implementation: implementing the plan within the existing administrative powers, ensuring the sources of revenue, and ensuring the availability of necessary personnel and responsible departments. The government must adjust any inconsistencies with existing laws and ordinances within the relevant departments, check the sources of revenue with the Local Finance and Budget Bureaus, and ascertain the availability of personnel and departments with the Administrative and Management bureau of the Ministry of Internal Affairs and Communications. While this step, 'adjustments required in organization,' is alluded to in the 'review of policy and plan draft' step of the five-step process, alternatives should be reviewed, including more realistic possibilities and new proposals related to the legal grounds, sources of revenue, and personnel. In actuality, alternatives generally get limited so as to fall within the administratively allowable range during a transportation strategy review, at the conference on a transportation plan for an urban area, so as not to include conditions that may be inconsistent with existing proposals for roads and railways. It is important, however, to make sure reviews can include new proposals that will increase the likelihood of reaching a truly effective and realizable transportation strategy proposal. In section 2.4.1 we will elaborate on this point.

Step (5), 'the social process of implementing the decision,' refers to a participatory approach toward decisions by involvement of a council, citizen participation, or a referendum, in order for a plan to be approved by a legislative assembly, thereby complementing the representative nature of the assembly. In general, the details of the proposal are discussed within the decision-making organizations, such as governors, mayors, in assemblies, community councils, and conferences, during each decision-making step. Although decision-making has been actively pursued step by step in

Table 2-1. The strategic planning process

Decision making process	Policy process	PDCA	PDS
Problem identification	Discovering the problem	Plan	Plan
Draft of a plan	The issue of choices among public/private sectors		
Review of the policy and draft, and the finalization of the plan	Problem solving techniques		
	Adjustment in organization		
	Consensus building		
Implementation	Execution	Do	Do
Evaluation and feedback	Evaluation	Check	See
	Feedback	Action	

these organizations throughout the construction of the transportation strategy, the importance of adjustment through external Public Involvement (PI) has recently been emphasized, so that more positive measures will be taken, as compared to the rolling on of the conventional process. Fig. 2-1 also highlights the importance of participation. It is necessary to introduce such participation from the initial step of the planning process not only to build consensus concerning the implementation of the plan but also to form agreements regarding the overall transportation strategy. In section 2.4.2 we will discuss this point.

Table 2-1 shows the relationships among the five-step process, the eight-step process, PDCA, and PDS. Problems can be found concerning the standards to be applied to feedback or review after action has been taken.

2.3 The Target of Transportation Strategy for City Regions

2.3.1 Historical Perspective

Examining the development of comprehensive transportation policy of the UK reveals that the focus of transportation strategy in urban areas extends beyond the reduction of congestion and air pollution to various other aspects of urban issues, such as overall management of the city and social inclusion, even though the axis remains transport. Principally, the role of a transportation plan was to provide transportation services to support various activities that contribute to the vitality of the city and to improve the quality of life. The setting of targets corresponds to the needs of the times; this adaptation is very important.

In Japan, the first person-trip research was conducted in the Hiroshima metropolitan area in 1967. It was recognized that there is a strong need for a comprehensive transportation plan in urban areas with populations of 300,000 or more, and so person-trip research is conducted essentially every 10 years, and mechanisms exist to reconstruct transportation planning for these urban areas. However, the necessity for re-planning is judged on an individual region basis, as part of the urban transportation administration within the City Bureau, and explanations are provided to the financial officials. Therefore, the implementation of re-planning is sometimes postponed or shelved, if agreement isn't established among the relevant organizations, such as the nation, the prefecture, the city, the towns and the villages. It is difficult for a public administration representative to understand the necessity for a long-range transportation strategy. The main problem seems to be a deeply rooted and fixed idea that if new roads or railways are not being proposed, plan drafting is useless. As already pointed out, the mental switchover has not been fully made from considering 'a plan' (a static thing) to undertaking 'planning' (a dynamic activity). There is inadequate understanding of the important roles of transport strategy and various policy considerations, as well as a failure to recognize that the target of the plan changes over time.

Paradoxically, the construction of a transportation strategy that contributes to the achievement of a policy target is important, even if these transportation facilities are not established at the present time. This is true, provided that the proposal for the new system to promote the transportation facility services and the management proposal to direct effective utilization of the existing transportation facilities has been drawn up in order to contribute to the achievement of the policy target.

Even if a plan has been carefully made, it will not be relevant forever. Review is required as time goes by, because the preconditions and the implementation environment will change. It has been pointed out that a flexible response is important to the strategic approach. Uncertainty regarding the plan environment, its value, and relationship-centered decision-making are not avoided in the drafting and implementation of the strategy. That is why the inevitability of a continuous review of strategy involving monitoring is emphasized. Even for long-range transportation plans looking ahead 20 to 50 years, regular reviews with approximately five-years cycles are necessary. This is because the planning environment as a precondition of the planning, the values associated with problem recognition and strategy decisions, and other relationship-centered decision-making issues will differ from the planning-centered ones determined at the beginning of the long-term planning period. In addition, quick response to unexpected change is also required. It is important to continuously and repeatedly conduct goal-setting activities, and the drafting and implementation of policies to contribute to regenerating city regions, while considering the overall goals and the limitations imposed by the economy, the environment, and the society, which today are indications of sustainability.

2.3.2 The Rearrangement of a Spatial Development Pattern

In long-term planning, rearrangement of a spatial development pattern is a key concept of sustainable land use/transport strategy. Turning to Germany, the Hamburger Verkehrsverbundes (HVV) was established in 1965 and proposed a railway corridor transport system (Fig. 2-2) to provide transportation service to support the city center. The transport system combines a rail network and access/egress roadway-based transport services, such as bus and park and ride. The spread of an Area-wide Parking Management (APM) system in Germany over the last 10 years has achieved good results, and has shown the effectiveness of this idea (Topp (1991)); (Retzko (1990)).

A basic problem for a city-center transportation plan is how to combine various modes of transport, including passenger and freight systems, buses, motorcycles, bicycles, and walkways. If the city's appeal is adversely affected by the concentration of automobiles over a certain level, it is effective to redistribute the street space in the city center by implementing comprehensive transport management (CTM) based on the 'city with people' concept, which enables people to walk with enjoyment (HMSO (1991)).

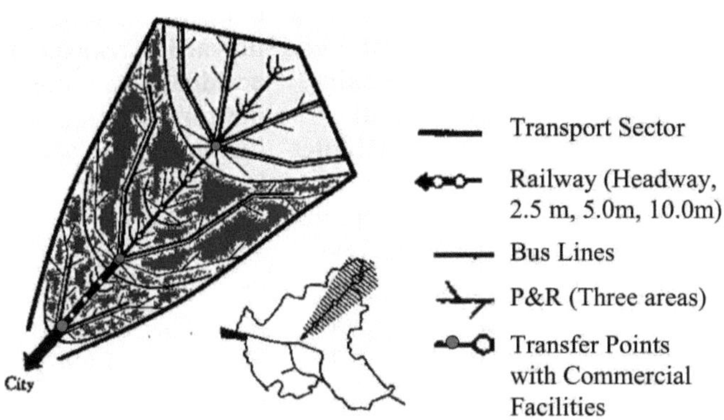

Fig. 2-2. Railway corridor transport system (Source: Hamburg Transport Association (HVV))

For example, pedestrian zones, fringe parking system and limited-period delivery permission can be incorporated in the CTM. Historic cities in Europe often introduce bus-based park-and-ride systems into the city center, in addition to implementing fringe parking. We must not forget that successful examples can be found representing various sizes and patterns of urban expansion, from small cities that permit the free use of automobiles to middle-sized cities that establish attractive pedestrian spaces with fringe parking system, to larger cities that implement a railway corridor transport system with Area-wide Parking Management.

These successful examples have formed the basis for many proposals, as well as comparisons among large cities concerning urban land use levels and transportation systems. Among these, our study group has focused on the proposal of intensive urban structures to integrate center systems, such as city centers or sub-centers, with public transportation networks, such as railways, LRT and BRT (Newman and Kenworthy (1999)); (Calthorpe (1999)).

After undertaking a comparative study of cities, UITP has pointed out: "Comparing social expenses per passenger, if the population density is over 20 people/ha, public transportation is more cost-effective than automobiles." (UITP (2002)) Table 2-2 show specific values for a representative city.

In the 1990s, Barter studied large cities of the world comparatively, focusing on Asian cities. Barter categorized the typical relationships among

Table 2-2. Population density and cost to the community of journeys (Source: UITP (2002))

Cities	Population density (inhabitant/hectare)	Share of journeys on foot, by bicycle and on public transport	Cost to the community of journey (as % of GDP)	Annual energy consumption per inhabitant (Mjoules per inhabitant)	Annual number of deaths in traffic accidents per million inhabitants	Access times by public transport in order to reach 500,000 jobs
Houston	9	5%	14.1%	86,000	130	70min
Melbourne	14	26%	11.8%	32,000	75	57min
London	59	51%	7.1%	14,500	35	31min
Paris	48	56%	6.7%	15,500	85	31min
Munich	56	60%	5.8%	17,500	55	26min
Tokyo	88	68%	5.0%	11,500	55	32min*1)
Hong-Kong	320	82%	5.0%	6,500	40	21min

1) Access time in order to reach 2 million jobs

city types, densities, and main mode of transportation (Barter (2000)). Although many large Asian cities are now classified as Bus Cities, Barter has described two possibilities of how they might deal with their future population growth. One is to become a transit city with the introduction of railways, LRT, and BRT, to develop mixed land use with high density around train/bus stations. The other is to become a high-mobility city dependent on bus networks with limited transport capacity, and with low-density suburban development dependent on automobiles. Barter insists that large cities should select the former approach.

Newman and Kenworthy have gone further and offered a conceptual diagram of the 'Future' Nodal/Information City, where public transportation is arranged for the development a city of generally low density with diffused nodes of higher density, so that the density of the nodes can be increased to permit an increase in the city's population. It is very interesting that the integrated development pattern with relatively higher density center development and a public transportation network have been suggested for both high-density cities in Asia and low-density cities in the US, Canada, and Australia.

2.4 Future Directions

2.4.1 Institutional Issues

One of the critical issues in urban transportation planning is to reform the use of automobiles and the automobile-dependent lifestyle that society has developed as a result of the historical accumulation of various influences. The influences of motorization have moved historically from possession of automobiles by a few to more opportunities for ownership and use, to the emergence of institutional locations that are dependant on automobiles, and on to the establishment of an automobile-dependent lifestyle. For example, in Japan at first the spread of automobiles occurred in parallel with population increases and involved the enlargement of cities. This enlargement led to increases in automobile ownership that exceeded population growth, and a much greater increase in vehicle kilometers traveled. Secondarily, institutional locations were promoted that depended on the spread of automobile use, and an automobile-dependent life-style has been established based on automobile-dependent land use and a high rate of automobiles diffusion. Finally, in order to support this historical trend, funding systems and land use regulation have been implemented, neither of which prevent increased demand for automobiles and suburban residential locations.

As described above, funding systems and land use regulations in an automobile-oriented society support institutional locations and lifestyles conditioned upon the increasing use of automobiles. If control of automobile use is demanded out of concern for sustainability, it is necessary to consider whether the expansion of the negative aspects of an automobile-oriented society are to be accepted in the planning system, and to provide the decision-maker with objective information about what effects will be brought about by the achievement of the policy targeted by the change. For example, as part of the planning changes required for city center regeneration, important factors include the cost effectiveness of land use regulations to transfer a prefectural office, city hall, or general hospital to the suburbs, or the new funding requirements for LRT construction. In addition, systems must be often be changed if pricing measures are introduced. However, price is a powerful force with the double effects of reducing demand for transportation while improving transport facility maintenance that will be covered by the resulting income. Thus pricing measures are expected to be introduced forcefully into transportation strategy.

In more general, local self-governing bodies must strengthen their ability to expand their revenue sources in order to escape from situations of easy dependency on passive development from the national government. They need to be able to switch to an active role in development whereby they can achieve more unique and locally appropriate solutions. A change in the law is required so that a governor or a mayor can suggest a tax system in order to ensure the necessary revenue sources, provided that the local people agree to the system and the governor or mayor is trying to promote a policy that meets local characteristics.

A mechanism might be introduced for helping a transport company that has run into the red and lost motivation to improve services (as can be found in some cities in Japan). Such a company, which provides the most cost-effective transport service to satisfy the conditions presented by a local self-governing body, could win a contract on condition of support funds being provided by the governing body, in the light of its role of providing transport service of a public nature. As this contract system would have a mechanism for evaluating achievement and the ability to reflect achievement through rewards, it would put pressure on the transport company to focus intensively on improvement of service, and as a result public transportation that suits the local characteristics could be improved more easily.

2.4.2 Consensus Building

This section strongly propounds that promoting the participation of the whole community in the transportation planning and decision-making process is required not only for obtaining agreement on plan implementation, but also for generating consensus on the transportation strategy targets. The process shown in Fig. 2-1 is expected to remove information gaps that often exists among the principal organizations, in accordance with certain local rules shared in common, and thereby enable the formulation of a common policy goal through the decision-making process. Typically, sufficient opportunities will be provided to adopt the opinions of various participants, and based on these opinions, the chief, the assembly, or the council will determine a transportation strategy. Without this kind of consensus, it will be difficult to actualize the policy proposals or planning system changes included in the strategic proposal.

As a specific example to show the importance of consensus formulation in transportation strategy, this section shows the difference between London, where road pricing was introduced, and Edinburgh, where a local referendum turned down the measures. After the measures were introduced in London, many communities in the UK and elsewhere recognized road pricing as a specific policy problem. In Great Britain, Edinburgh would have been the second city to introduce these measures, and it was expected they would be approved by the local referendum.

In actuality, however, the referendum (conducted in the form of a mailing executed from February 7 to 21 in 2005) forced an end to the plans for introducing tolls, because three-fourths (179,905) of the voters were overwhelmingly against the measure (there were 45,965 in favor). The planning draft proposed that each automobile going into the center (during the hours of 7:00 to 18:30) or to the city periphery (from 7:00 to 10:00) would be required to pay two British pounds each day they made the trip, and it was explained that the profits would be allotted to investments in transportation: specific measures were drawn up for trams and airport railways and the expansion of the 20 mile transportation zone (Transport Edinburgh Referendum Archive); (The City of Edinburgh Council 2004).

What was the difference between the two cities? One clear difference was that in London, road pricing had been included among election promises during the mayoral election, and they were promoted after the candidate won, while in Edinburgh, the comparison of strategies based on the presence or absence of road pricing was entrusted to the proponents and opponents of the local referendum. The measures in both London and Edinburgh required a change in current actions and financial burdens, and were presented as part of a packaged measure, and they were similar in that the transportation strategies had been approved by the local assemblies via public hearings. But then London implemented public consultation without incurring any legal obligations concerning the Congestion Charging Order, determined the contents of the plan, and executed it. Edinburgh, on the other hand, held a public hearing under the Congestion Charging Order, but after approval by the assembly, implemented the referendum in order to respond to a request from the Scottish government "to show the clear support of the citizens," which forced the cancellation.

This fact indicates that a measure imposing a new financial burden involving a change of action is unpleasant for citizens to accept, even though the beneficial effects of the additional measure have been explained. In London, it was reported that an opinion poll before introduction of road pricing showed that 60% of the people were against the measures, but in the current situation with a 20% reduction in traffic volume, the number of people for it increased to 60%. The important factors leading to the introduction of road pricing were as follows: even though it was not a popular policy, but it was socially necessary, it was presented as an essential proposal for achieving the policy target; and it was introduced carefully but all-out so that the effects could be seen quickly.

Another factor was that Edinburgh had sufficient transport revenue. Thus, a clear relationship did not exist between revenue and maintenance, such that the revenue obtained by the introduction of road tolls was seen as essential to maintenance (Byrd 2005). In addition, it has been pointed out that just before the referendum, the declaration that "the Scottish government will construct two tram rails independent of road tolls" had an effect on the results of the vote (Goodwin 2005). These factors show that it was problematic to discuss the road tolls to be included in a financial package with the maintenance of the two tram rails.

In Japan, recently, the candidates in an election typically contend with one another by presenting their manifestos as election promises. In particular, during a mayoral election in Sapporo City, in Hokkaido Prefecture, the construction of an underpass for Sosei River Street and re-maintenance including underground passages in front of Sapporo Station were big issues, although they already had been determined as community enterprises. The candidate who pledged to gather opinions from the citizens and review those enterprises with reference to the opinions won the election, and then took measures including a workshop with 1,000 participants. In the workshop, the basic target was conveying the transport vision for the city center and obtaining either approval or disapproval of the two major projects (City of Sapporo 2004). The results of the workshop were presented to an expert committee. After the committee completed a transportation plan for the city center and presented it to the assembly, the assembly adopted the plan by a majority vote. The clear positioning of the workshop with its 1,000 participants can be seen as very important to the process of drafting and deciding on the transportation plan for the city center. In addition, the following two factors contributed to formulating a consensus: the review results of the effects and the sources of revenue were fully provided as data for the workshop and the expert committee, and they were applied as support information for the decision-making process. It is also very important that this decision-making support information was appropriately presented during the decision-making process leading to the decision made, regarding the issue that the mayor had raised.

2.4.3 Local Governance

The relationship between the administration and the citizens in a local autonomous area is certainly not a negligible factor. The time when an administration reigned over people unilaterally has passed, and now we are in an era when an administration governs cooperatively with the citizens. The shifts from government construction to local governance, or from passive development to positive development, are often pointed out.

A local autonomous area is said to solve problems on the local level, decide local policies and establish local rules, in a style different from that of the nation, in that it does not play a role in promoting governmental policies as they are, as a branch office of the central government. A new relationship between the administration and people has been constructed in order to bring about the objective of local autonomy, under the social change of decentralization of the political system, transparency (information disclosure and accountability), and the principle of adherence to administrative procedures.

Strong resistance is met when approval or disapproval based on citizen participation and local referendum results are provided as important inputs, because some have a negative opinion of the current system that implies people's opinions are useless, and some have positive opinions about the current system and think that it is sufficient for the assembly to decide under the principles of parliamentary democracy. These negative opinions can be thought of as being deep-rooted in the rivalry between people and administration, because the administration believes that people only make requests. Concerning these opinions, many researchers have pointed out that it was natural that residents became demanding under a system where service levels changes became residents' burdens without reflection or the provision of necessary information.

In addition, as a further contribution, Muto has explained the appropriateness of citizen participation in councils and decision-making means such as referendums as follows: "The reason the assembly has decision-making authority is that it is the representative organization of the nation's citizens or residents, but the opportunities to check the assembly comes in the form of elections held once every few years (Muto 2003). Therefore, it is necessary to improve the representative function of the assembly by means of compensations in the representative quality of the assembly and by methods other than elections, so the nation's citizens/residents can participate in the decision-making process, or can decide directly."

Some have talked about a 'complementary principle' in local autonomous areas, which has recently been focused on, and has been summarized in terms of resident self-governance. Majima explains that the complementary principle refers to the "distribution principle: public problems and business should be divided up and assigned, according to priority, to cities, towns and villages as basic self-governing bodies; if they can not solve the problems, prefectures (as large regional self-governing bodies) should take charge; only after that, if the problems still remain, should the central government handle them." (Majima (2003)) He has also pointed out that, from the perspective of resident self-governance, "First, public problems should be solved at the individual citizen level through self-governing activities, if a citizen can solve it; if the problems cannot be solved at the citizen level, they should be solved at the community level; only after that, if the problems still remain, are the problems to be seen as administrative issues for the cities/towns/villages; the administration will try to solve such problems with citizen participation and consultation."

It is certain that awareness that 'there are no easy ways in public administration' has spread among grass-roots movements in Japan. Even in relation to transportation, the activities of citizen groups, including RACDA and LRT Sapporo, are beginning to be noticeable. This 'complementary principle' may the natural trend for mature local self-governance. In this way, even in times when a personality as a sovereign is strongly expressed, that sovereign will presumably be required to provide information to the citizens as well as the administration, for example as a decision-maker regarding transportation. In a highly transparent decision-making process, understanding new needs, a sovereign must provide scientific and objective information that is as easy-to-understand as possible.

Citizens are also 'customers' of administrative service. From the customer's point of view, it has been pointed out that "the focus must include the problem of relations between the benefits and burdens of administrative service, the problem of ensuring efficiency when providing services, the problem of evaluating achievements, and the management the issues of privatization and service-providing systems." The presentation of scientific and objective information to meet these interests is an important issue when implementing transportation strategies. In closing, the author hopes that the contents of this essay will be an aid to thinking about the improvement of transportation strategies in city regions.

References

Barter, P.A. (2000) "Transport Dilemmas in Dense Urban Areas: Examples from Eastern Asia", in Jenks, M. and Burges, R. (eds) *Compact Cities: Sustainable Urban Forms for Developing Countries.* Spon Press

Byrd, T. (2005) "Edinburgh hassle", *Transportation Professional*, March 2005

Calthorpe, P. (1999) *The Next American Metropolis; Ecology, Community, and the American Dream.* Princeton Architectural Press

City of Sapporo (2004) *Guidance of community-based transport planning* (in Japanese)

Goodwin, P. (2005) "Little good to learn from tale of two cities", *Local Transport Today*, 415

Harata, N. (2003) "Transport Strategy in Asian High-density Metropolitan Area" in *A Research Report (A-342) on Comprehensive Transport Planning* (issued by the Japan Research Center for Transport Policy) (in Japanese)

HMSO (1991) "Chapter 18: Appraising Trunk Road Schemes in Urban Areas" in *Volume 12: Traffic Appraisal Manual*, August 1991, in *Design Manual for Roads and Bridges*, May 1996

LRT Sapporo (2004) *LRT can change Sapporo: A LRT10km plan in Sapporo* (in Japanese)

Majima, M. (2003) "Autonomy right expansion of small community" in Okamoto, Y (ed) *Theory and Practices of drafting a Policy.* Hosei University Press (in Japanese)

Muto, H. (2003) "Chapter 3: Basics of Policy Process" in Okamoto, Y. (ed) *op. cit*

Newman, P. and Kenworthy, J. (1999) *Sustainability and Cities.* ISLAND Press

Ohta, K. (1988) *Transport System Planning, Japan Society of Traffic Engineering* (in Japanese)

Retzko, H.G. (1990) *Lecture on parking policies in Germany cities* (Tokai Club in the Kasumigaseki Building in Tokyo, July 24, 1990)

The City of Edinburgh Council (2004) *Local Transport Strategy 2004-2007*

Topp, H.H. (1991) "Parking policies in large cities in Germany", *Transportation*, 18

UITP (2002) "Public transport for sustainable mobility", *Focus* (A UITP Position Paper), July 2002

Transport Edinburgh Referendum Archive, http://www.edinburgh.gov.uk/transportedingurgh/Referendum/index.html (Accessed in 2007)

Part II

CHINA
KOREA
INDONESIA
THAILAND

3. Governing Urban Development in Dualistic Societies: A Case Study of the Urban Region of Yogyakarta, Indonesia

Leksono Probo Subanu

3.1 Introduction

The development of cities in the developing countries often takes on a character of a certain mixture between modernity and tradition. This mixture can be seen in the mixture between structured urban forms of the modern sector and organic landscape of the traditional compounds. Socially, the mixed or dualistic nature is also evident in the public behavior of urban dwellers, which is only partially based on formal rules of behavior while traditional customs for social relations often still dominate. The nature of duality may well explain some behaviors in formal institutions, where written laws and rules more often than sometimes are not properly observed and enforced. The incidence of official corruption, for example, may have been partially influenced by the generally slack observance of rules and regulations that characterizes dualistic societies.

The sustainability of an urban region depends upon the ability of urban communities to sustain sufficient economic activities while containing their externalities on the environment. Local governments, community organizations, and other stakeholders of the urban region need to have sufficient capacities to mobilize and organize social, economic, natural, and institutional capital within the urban region to achieve those ends. It is the responsibility of local governments, which have the legal powers to initiate and implement public policies and actions for the good of all citizens, to take the role of leader and facilitator for the efforts towards the sustainability of the urban region. However, when an urban region is composed of several local government entities, the seemingly insurmountable problem of integrating public policies and programs among the local governments emerges. While in many urban regions in the world, institutional and legal solutions to this problem have become common practice, it is often still an intractable problem for countries in the developing world. This paper intends to highlight the problems faced by three local governments in managing urban development in the greater Yogyakarta urban region. It will elaborate on the effects of duality in the implementation of policies related to urban development.

3.2 Yogyakarta in History

The historical background of the city dates back more than a thousand years to the Buddhist dynasty of Syailendra, which in the 8th century erected the famed temple of Borobudur to the northwest of the present location of the city of Yogyakarta. Closer, toward the east, are a large number of temple complexes erected in later centuries, the largest of which is the temple complex of Prambanan, which was erected by the Hindu king Dhaksa of the ancient Mataram kingdom in the 9th century. This tradition of ancient empires in the vicinity of the present location of the city or Yogyakarta continued with the return of the later Mataram kingdom in the 16th century to the area, after a period of several centuries during which the center of power on the island of Java shifted to the eastern part of Java and subsequently to the northern coast of Java with the entry of Islam. The new Mataram dynasty was centered at an area called Kotagede, which is now part of the city of Yogyakarta proper. After the passing of Sultan Agung in the mid-17th century, during whose reign the Mataram kingdom became very influential and sent an army in a failed attempt to oust the Dutch colonial government in Batavia (at present, Jakarta), the kingdom went into a period of internal conflict and struggle for dominance among members of the dynasty. This period of weakness offered an opportunity for the Dutch colonial government to intervene. Under the influence of the colonialists policy of 'divide and rule,' in the middle of the 18th century the kingdom of Mataram was finally split into two kingdoms, the Sultanate of Yogyakarta and the Sunanate of Surakarta, located about 60 km to the northeast. Thus, twin dynasties of Yogyakarta and Surakarta were created and continue to exist until today.

The division the city of Yogyakarta was started in the mid-18th century by the Sultan Hamungkubuwono I, by building his palace in 1756 on a well-chosen site between two rivers, Code river and Winongo, in Beringin forest. The Sultan chose a site for his palace not far from the site of the previous Mataram palace in Kotagede, about equidistant from the Merapi volcano and the south coast. This choice created a long-standing legend that the Sultanate is in some supernatural way related to the unseen ruler of the Merapi volcano and the so-called Queen of the South Seas, which many locals still believe to be true. Objectively, the choice of Sultan Hamengkubuwono I was quite wise. The site is about 25 km from the foot of the volcano and is protected by two minor peaks at the foot of the volcano and deep gorges that contain lava and volcanic debris flowing from eruptions. The site's elevation is about 100 m above sea level and about 25 km away from the coast, evading natural hazards coming from the sea such as severe storms and tsunamis. The abundance of water in the surrounding countryside guaranteed the development of agriculture that would supply food for the new kingdom.

The structure of the city that emerged after the palace was built was determined by a number of historical events. First, the Sultan granted land to people who worked in the palace in the surroundings of the palace. The workers were grouped according to their specific occupations, such as musicians, soldiers, gamelan players, iron smiths, etc. These compounds were named after their occupations and today they retain their original names, even though the inhabitants no longer work in the palace. The Sultan also granted lands farther away from the palace to his high officials, on which each official built his house and in turn let his workers live in the surrounding area. This pattern remains today in areas around the palace up to about 1 km away from it. These official houses became the nuclei of what today is called the kampong located in palace vicinities.

Second, the Dutch colonial government set up a colonial government to oversee the Sultan's activities. The colonial government built a governor's mansion and a military fort on both sides of the road guarding the entrance to the Sultan's palace, thus effectively controlling the activities of the Sultan. The colonial government further encouraged the Chinese traders to occupy an area close to the colonial government site to set up businesses. Thus, the central business district of the city was established on the street connecting the palace complex to the rail station to the north.

Third, the short British colonial rule at the beginning of the 19th century managed to divide the Sultan's influence further by establishing a rival, albeit lower, princedom of Pakualam. This division lasted until after the nation's independence in 1945, as both sides voluntarily merged to form a single local government at the present provincial level.

Fourth, after independence, the Sultan Hamengkubuwono the Ninth, being university educated and thus eager to promote higher education, granted a significant piece of land of more than 100 hectares to Gadjah Mada University, established in 1949. The land was situated in the northern outskirts of the city, effectively outside the administrative boundary of the city and thus under the administrative jurisdiction of Sleman regency.

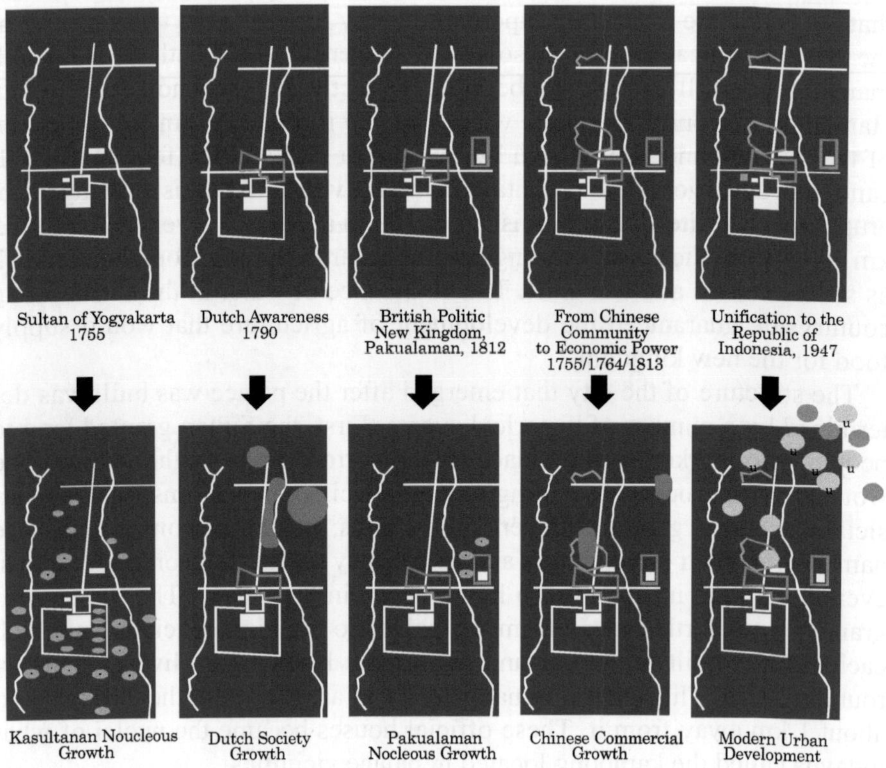

Fig. 3-1. Sketch of historical development of the Yogyakarta urban region (Used with permission of Laretna T. Adhisakti)

In the 6 decades since its inception the University has grown to become the biggest and most comprehensive higher education institution in Indonesia. It became a sustaining factor for the development of the entire urban region, especially during the era of the Indonesian republic.

3.3 Yogyakarta at Present: Sustainability in Question

There are actually two primary centers of development in the Yogyakarta urban region, namely the Kraton or Sultan's palace with its traditional cultural and commercial activities and Gadjah Mada University with its educational activities. The two centers are located relatively close to each other, the Kraton at the old center of the city and the university at the northern periphery about 4 km away.

The development of higher educational activities during the past 5 decades shows the significant influence of education on the development of the urban region. Gadjah Mada University was the first national university to be established in Indonesia. Since its establishment, the university had attracted students from all over the country, making it a sort of a melting pot of the much-varied Indonesian cultures. The city came to represent the variety of ethnicities and cultures that exist within the entire nation. As the university grew over the years, it sparked the emergence and growth of other higher education institutions, national and private alike. Specialized national institutions such as the arts institute, teachers' institute, and Islamic institute were established. Various private colleges that later became full universities as well as specialized private institutions grew rapidly since the 1960s and 1970s. The number of academic institutions eventually grew to more than 80 at its peak with more than 290,000 students making up a significant proportion of the entire population of the urban region. Educational activities in academic campuses and the students' living accommodations outside the campuses[1] thus contribute a very significant proportion of the local and regional economy.

The significant number of students coming from other areas of the country increases the demand for housing facilities in the urban region. Student housing needs became the defining factor in economic development. Many households were converted to become boarding houses for the students. Many parents from outside the area decided to buy houses in Yogyakarta to accommodate their children rather than to rent rooms. This drove the housing development industry to high gear, especially during the 1980s and 1990s. The demand for land for housing development drove the price of land near the academic campuses to the sky. For example, land price in an area about 1 km away from Gadjah Mada University was driven more than 100-fold in about 20 years.

The old center of the city, namely the Kraton or palace with the surrounding traditional cultural and commercial spaces, in the meanwhile grew to be a center of touristic attraction. The Kraton itself is a very popular museum piece for domestic as well as foreign tourists. The batik industry located south of the Kraton developed to be a very popular lodging place for tourists. Kotagede, the original location of the kingdom of Mataram, developed as a center for artisans and silversmiths, attracting visitors to buy articles and to watch silversmiths at work. The unique spatial pattern of houses in the Kotagede area also became an attraction for visitors and investors. The craft industry became another strong driver for economic development. The local social and human capital that exist in the population are conducive to the development of craft industry, making Yogyakarta a strong exporter of handicrafts and artistic articles.

Because of these tourist attractions and the internationally significant archeological sites of Borobudur and Prambanan located close to the city, Yogyakarta became the hub of tourism on the island of Java. A good number of private investors from outside as well as local investors laid their capital in developing tourist accommodations. Tourism thus provided the impetus for economic development in the city and the region around it.

The above shows that primarily there are two axes for economic development of the urban region of Yogyakarta, i.e. higher education and tourism. The development of the regional economy in the 1980s and 1990s was largely driven by the two axes. However, recent developments in the national and international scene have changed the situation. The economic crisis of 1998 has caused a prolonged slowdown in the national economy that has affected the flow of visitors to Yogyakarta. Terrorist acts in Indonesia and outside the country since the start of the 21st century have had debilitating effects on domestic as well as foreign tourism activities. As in other tourism dependent economies, such as Bali, the economy of Yogyakarta was heavily hit by these crises.

The other axis for economic development in Yogyakarta, i.e. higher education, was also affected by the economic crises, especially in terms of the post-1998 slowdown of capital investment. Additionally, the competitive advantage of Yogyakarta in higher education services was being eroded by the emergence of universities in almost all provinces in Indonesia, while the number of students graduating from high school has begun to decrease significantly, thanks to the success of family planning programs. It is also evident that the reputation of Yogyakarta as a good place to send children to get higher education has been somewhat eroded because of the increasing drug abuse and loosening of sexual mores among the students in Yogyakarta. This seems to be the result of the shifting pattern of student residence from family-based rooming in which students become members of the family to the rent-based boarding houses where students are on their own. These events drove down the number applicants to higher education institutions in Yogyakarta, forcing some of the weaker ones to go out business or to merge.

At the present time the situation of both axes of economic development in Yogyakarta urban region is still in a challenging if not difficult state. The significant decrease in tourism and the continuing decline in student enrollment in universities challenge the region to find alternative ways to support economic development in the future. The recent earthquake that shook the region and destroyed more than 200,000 homes, damaging 80% of productive assets in the area, mostly small enterprises (Java Reconstruction Fund 2007), added to the region's economic development challenges the need to restart small enterprises and restore the industrial economy after the quake.

These challenges to the economic base of the Yogyakarta urban region introduce a certain level of vulnerability to the sustainability of urban development and regeneration. If the decline of the economic base could not be turned around, urban regeneration would be jeopardized. The local and provincial governments are currently working to overcome the problems. Unfortunately this is not the only challenge to the sustainability of the urban region. The spatial development of the urban region also poses a number of problems that challenge sustainability. The following describes the problems of spatial development in Yogyakarta.

3.4 Spatial Development of Yogyakarta

The historical account of Yogyakarta shows that the city evolved from a single nucleus for about 200 years since its inception into a dual-nuclei city with the development of Gadjah Mada University as the second center of development. The proximity of the two centers of development, however, has in effect merged them as a single center with two development characteristics or axes, i.e. tourism based and higher education based.

During the economic boom of the 1980s and 1990s, these two axes attracted investments in property development primarily in housing and commercial facilities. Commercial, especially retail and office facilities, sprung up along the main roads in the city and gradually crept over the main thoroughfares leading out of the city proper. New college and university campuses were built and old campuses were renovated and upgraded with multi-story buildings. Houses in areas surrounding the campuses were converted into rooming houses for students, often to the effect of kicking out the original inhabitants, while also making the areas more densely built up.

The above developments within the city of Yogyakarta created the effect of out-migration from the city toward the outlying areas located in Bantul and Sleman Regencies north and north of the city, strengthening the urbanization of the Yogyakarta urban region. Fig. 3-2 and Fig. 3-3 show very clearly the trends of population growth in the city of Yogyakarta and the neighboring areas in the last 20 years as measured from formal census data. It can be seen that the city of Yogyakarta has stagnated or even lost some of its population, while certain districts in Sleman and Bantul regencies at the periphery of the city are gaining population very rapidly. This data suggests that in the past 20 years, people are moving out of the city towards the peripheral areas. It can be observed that the new peripheral housing was populated by starting families seeking a place to live, established families moving out of houses that were converted to other uses in the city, and families coming from elsewhere for retirement or study. Many of the new houses are bought for investment, not for dwelling, as is common in other parts of the country, especially Jakarta, where apartments are bought up by people who are not in need of dwelling, but are putting the apartments up for rent or for eventual sale for profit. All these factors drove up the demand for housing in the Yogyakarta urban region, which in turn attracted many housing developers.

There are two kinds of housing developers. One kind are those developers who formed formal corporations that go through the formal permit, construction, marketing and sales process and thus can be easily controlled by the local government. These developers usually acquire open agricultural lands, apply for conversion to housing status, and build from about 50 to hundreds of houses. The other kind of developers are the small informal developers who bought plots of housing land, primarily within existing rural settlements which have relatively large plots, subdivide them and build houses for sale. This type of developers work below board, almost invisible, and outside the development control system. It has proven very difficult to control the activities of the small developers, as the local

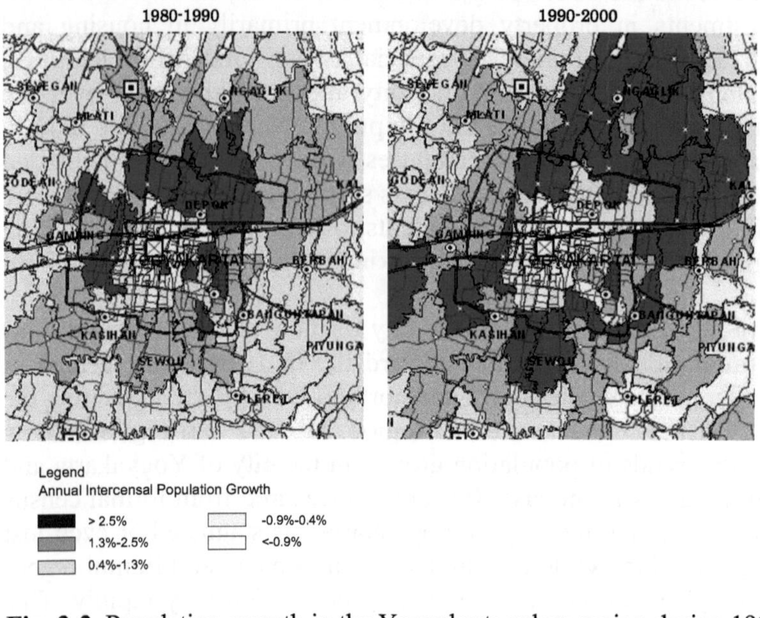

Fig. 3-2. Population growth in the Yogyakarta urban region during 1980-1990 and 1990-2000 (by sub-district) (Base map courtesy of PSPPR-UGM)

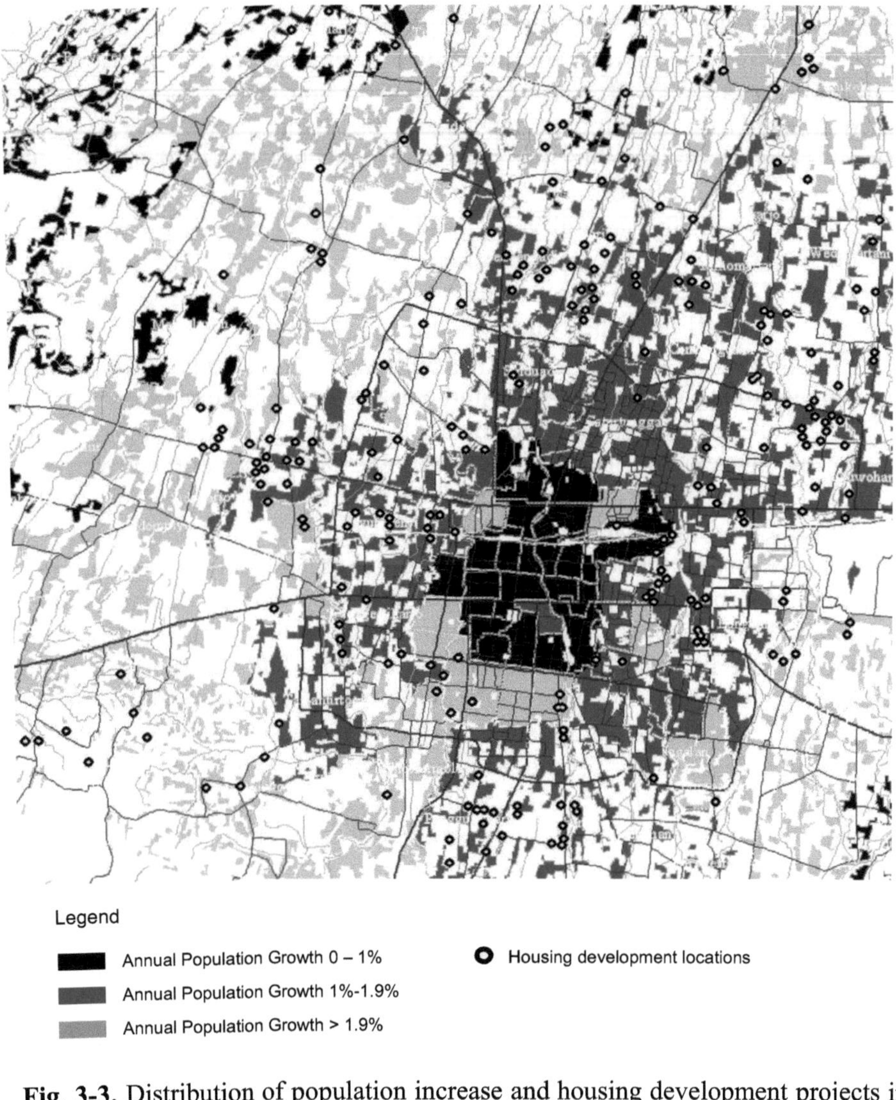

Fig. 3-3. Distribution of population increase and housing development projects in Jogjakarta (2005) (Used with permission of Kartamantul Joint Secretariat Office)

governments do not have sufficient abilities to monitor the development activities, much less to control them (Syarifuddin (2000)); (Srianto (2000)).

The result of this situation is evident in the densification of the rural housing settlements and the rapid conversion of agricultural lands into housing developments. Spatially, the conversion of agricultural lands occurred in a haphazard fashion with no apparent pattern. The formal developers usually start their projects by acquiring land. The acquisition of land did not follow any spatial pattern or plan at all. Basically, the developer will look for landowners who would sell their land at a good pricc --- i.e. price at which the developers can have a profitable margin catering to a certain housing market, buy the land and then built and sell houses. The more moneyed developers would buy land at cheaper prices in outlying areas and then wait for several years until the market caught up with them, and then start the housing construction. All this resulted in an uncontrollable sprawl of housing developments with from tens to hundreds of houses in them spread all over the countryside. Fig. 3-3 above shows the housing developments that seemed to be randomly spread over the rural agricultural areas among the paddy fields and existing rural settlements.

The local government has the means to control such housing developments, but the existing regulations governing land development and construction are generally unable to achieve effective control. Many studies have focused on the development permit system in a number of local governments. Most of them found that the development permit regulation itself have not been fully observed by those who started construction of buildings. Many people built their houses without applying for a building permit and got away with it without any sanctions. The common practice is for local governments to periodically give pardon for those who violate the rule, so that the rate of compliance can be increased. (This means the collection of revenue from people who are pardoned, and a good record for the department administering the permit system.) This situation makes the development permit system an ineffective tool for control, because there is no punishment for violating the law.

With respect to the effectiveness of the development permit system to influence the location of development and economic activities, there is the problem of inconsistencies between the substance of the spatial plan and the granting of permits. The substance of the spatial plan developed in local governments generally are local master plans based on a scale of 1:5,000, which is generally not sufficiently detailed to be a reference for permit granting decisions for individual parcels of land. Therefore, the permit-granting officials would generally use their own interpretation for deciding whether or not to grant the permit. For example, the question of how much of a block of dwellings should be granted a permit for converting into commercial use is up to the permit-granting official. In many instances, landowners build or convert their existing buildings without bothering to apply for a permit, and most often they can get away with it if the building or the conversion is relatively small and unobtrusive to the general landscape. If this was allowed to go on for a prolonged time, suddenly everyone will find that a whole block of dwellings would be converted into other uses. The eventual effect is obvious: the spatial plan is generally ignored. In the case of the Yogyakarta urban region, the ineffectiveness of spatial plans was exemplified by the unplanned and pattern-less conversion of agricultural lands into housing settlements, as well as the densification of rural settlements in the peripheral areas of the city of Yogyakarta.

3.5 Spatial Planning Institutions

Spatial planning as a formal government responsibility was established in 1980 with the issuance of Ministerial Decree no. 4/1980. This decree was issued in lieu of a national spatial planning law that was deadlocked on the drawing board for almost a decade. After the ministerial decree, all cities (municipalities) and towns (unincorporated regency towns) were obliged to have spatial plans. According to the decree, spatial plans for cities and urbanized areas should be formulated at three levels, i.e. the general plan (scale 1:10,000 to 1:20,000), the detailed plan (scale 1:5,000), and the technical plan (scale 1:1,000). Spatial plans for regencies (regional local governments) were to be general plans at the scale of 1:100,000 to 1:250,000 depending on the size of the regency. The central government provided grants to local governments to fund spatial planning projects for about a decade after the inception of the decree[3].

A spatial planning law was enacted in 1992, after the deadlock between the Ministry of Internal Affairs and the Ministry of Public Works was successfully broken under the leadership of the Ministerial Office of Environment[4]. Law no. 24/1992 on Spatial Planning was basically a mandate for local and provincial governments to institute spatial planning as part of their mandatory activities. It contained provisions on who-does-what in the formulation of spatial planning. The hierarchy of spatial plans from the national, provincial, to the local levels was also stipulated. It is unfortunate, however, that the participation of the citizen had not been very explicitly stipulated in the law. Also unfortunate that the relation between spatial plans and implementation mechanisms at the local level was not dealt with in the law. Thus, Law no. 24/1992 on Spatial Planning was basically a plan-making law, and did not address sufficiently the creation of a planning system that incorporates implementation at the street level. Implementation of spatial plans was not very clearly defined in the law. Institution and tools for implementing spatial plans at the local level were not defined at all.

Law no. 24/1992 on Spatial Planning in effect formalized the institutionalization of spatial planning in the local and provincial government.

Thus, local governments are obliged to formulate spatial plans, but lack the mandate and impetus to create institutions and regulations to implement them. Generally local governments use the existing development control regulations, sometimes too loosely, to enforce the spatial plans. The existing tools include the building permit (for new construction and renovations), location permit (for development of certain facilities of certain size), and business permit. According to some studies, the building permit and location permit regulations have shown some effectiveness in controlling the use of land (Suryati (2003)), even though in many instances the permit regulations had not been explicitly linked to the spatial plans. The business permit, which was originally a tool to levy revenues from starting businesses, has rarely been used for land use control. It would have been an effective tool for making sure that conversion of land uses would comply with spatial plans. The lack of effective development control mechanisms at the local level resulted in the low effectiveness of spatial plans to influence actual developments.

Recently in March 2007 the law on spatial planning was replaced by the new Law no. 26/2007 on Spatial Planning. The new law was initiated in the Ministry of Public Works, collaborating with the Ministry of Home Affairs and other relevant ministries. The new law contained more detailed description of the types of plans for levels of government, the hierarchy of the plans, the role of the people in the planning process, some very technical provisions on the amount of green space in cities, additional provisions on sanctions, and some mention of mechanisms such as incentive-disincentive and zoning. However, mandates on implementation institutions and regulations are not clear; the people are not included in planning decisions (they are just allowed to give inputs), and as many as 16 important provisions in the law are delegated to bylaws. Thus the new law is still basically a mandate for plan formulation which lacks clear provisions on implementation. The writers of the law seem to take the existing implementation tools and institutions for granted.

The urban region of Yogyakarta has not been exempt from the above conditions. Management of urban development for the urban region is the responsibility of three local governments, namely the city of Yogyakarta, the regency of Sleman, and the regency of Bantul. Each local government has its own system for controlling land and spatial development. There are similarities in terms of the kinds of institutions responsible for administering the building, permit, and business regulations, but the policies and attitudes behind the permit granting process may be different.

It interesting to mention that Sleman regency, ahead of other regencies, has initiated a groundbreaking regulation that attempts to control the conversion of rice fields into housing tracts, a measure directed at conserving rice production on fertile and irrigated fields and directing housing development to less fertile and productive lands. This regulation seems to have some effect at steering housing development away from fertile rice fields. However, the effect seems to be limited to the activities of registered developers who work above board. The small, agile, and practically invisible developers who work below board remain difficult to control. Bantul regency has not adopted new regulations to control housing development (Urban and Regional Planning Program, Gadjah Mada University (2007)). Instead, the Bupati (head of local government) in Bantul personally oversees the development permit process for housing developments to prevent fertile rice fields from being converted into housing tracts. In the city of Yogyakarta, the Mayor has put all permit granting processes (including building permit, location permit, and business permit) under a single agency that is responsible for permit services. This agency will coordinate with the technical and substantive agencies for the evaluation of the permit applications.

Such differences in approach and method for controlling development among three neighboring local governments is obviously not supportive of the need to manage the contiguous urban region of Yogyakarta in an integrated manner. However, there is yet no legal mandate for the handling of such situations, so the three local governments carry on according to their own perceptions of a common problem and find their own solutions separately. This is a problem in intergovernmental policies and actions that is to a large degree affected by the decentralization policies of the central government.

3.6 Decentralization and Its Consequences

Since the downfall of the strongly centralist regime of President Suharto, cries for more principled decentralization became louder. The old Law no. 5/1974 on Local Governance was considered to be too easily undermined by centralizing policies, where local governments were almost completely dependent on central government grants for recurrent as well as development funding. Law no. 22/1999 on Local Governance and Law no. 25/1999 on General Revenues were put into effect by President Abdulrachman Wahid in the year 2000. These laws delegated all governmental affairs to the local governments (except for religious affairs, foreign affairs, the military, monetary affairs, and justice) and provided formulas for revenue sharing in the form of unspecified transfers from the central to the local governments. In five years these laws were replaced respectively by

Law no. 32/2004 on Local Governance and Law no. 33/2004 on General Revenues. The new laws provided the necessary changes based on lessons learned during the 5-year application of the old local governance law.

The most significant effect of the new decentralization laws was that local governments became much more independent than previously. The law also allows local government territorial units to be divided into two or more units of local government. The effect was tremendous. Within 7 years the 27 provinces became 35 provinces and 226 local governments multiplied into more than 400 municipalities and regencies. Local governments became very much inward-looking and self-centered. Each tries its hardest to increase local revenues, so much so that it became very greedy towards all kinds of business and in effect detrimental to the investment climate. (More than 200 new local legislations on local taxations were subjected to review by the Ministry of Home Affairs for unjustified revenue generation provisions.)

The decentralization law made it possible for local governments to set up their own institutional arrangements within the limits set by the law. This new freedom can actually be seen as an opportunity to make innovative institutional arrangements for the local governments. However, discretion for policy making without sufficient capacities often lead to excesses. For example, a regency located in the middle of the jungle decided to build a theme park. Another built a government complex that was so extravagant that the regency will be in debt for years to come. Some of the more prudent local governments adopt useful innovations, such as Sleman regency with its regulation and agency for controlling conversion of agricultural land.

3.7 Inter-governmental Collaboration

Opportunities for horizontal collaboration among local governments are provided by the law. However, not many local governments are as yet inclined to collaborate with their neighbors. The Yogyakarta urban region in this case has been a nationally recognized innovator. The establishment of a joint secretariat among the city of Yogyakarta and the regencies of Sleman and Bantul to facilitate coordination for collaborative cross-border policies and programs was a first in the country. The Joint Secretariat was established in the year 2000 after 8 years of preparations and persuasion by the provincial government. At this time, the Joint Secretariat's activities are focused on 5 areas of infrastructure and services, i.e. solid waste management, sewerage and sewage treatment, rain water drainage, public transportation, and clean water. The function of the Joint Secretariat at the present time is that of facilitation and mediation for policy coordination and integration among the three local governments in the respective areas of infrastructure and services. Decisions for development policies, programs and projects remain with each local government. Thus, the success of the Joint Secretariat actually depends on the ability of each local government's officials who are members of the Joint Secretariat to convince each respective local government's policy makers to adopt and follow through policies recommended by the Joint Secretariat, and the commitment of the policy makers in each local government to the joint effort. After several years of support by a GTZ project from Germany, in the last 3 years the Joint Secretariat is funded by contributions from the three local governments. This shows that there is a high level of commitment of the three local governments establishments to use the Joint Secretariat as policy making tool. This includes the local parliaments, which are generally very much inward looking.

It needs to be said here that the innovation of the Joint Secretariat was made possible by the insistence of the Provincial government to put the three local governments together to deal with cross-border issues in the Yogyakarta urban region. The effort was started at the beginning of the 1990s with the entry of the Swiss-funded Yogyakarta Urban Development Project. The project quickly found the need to integrate policies in the three local governments to deal with problems in the Yogyakarta urban region. Working through the Provincial government, the project sponsors held meetings and discussions among officials of the three local governments. The Provincial government was strongly supportive of this effort. The Province of Yogyakarta has only 5 local governments (1 municipality and 4 regencies) to supervise, which explains why it was able to provide strong support for this venture. Furthermore, the Sultan as Governor of the province has much more persuasive power over the local governments, which in the old kingdom days were his subjects, than do other governors in the country. This is clearly a legacy of the old Mataram kingdom, that is the unity between the subjects and the ruler (Javanese: *manunggaling kawulo lan gusti*). When the Joint Secretariat was finally established in 2000 through a joint decree among the local governments, the Governor/Sultan also signed the decree to show his support.

Recently, the Joint Secretariat started a new venture into collaborative spatial planning. This can be seen as a big step forward. Previously, the work of the Joint Secretariat was basically coordinating infrastructure development which are concrete objects with problems that need immediate solutions, while spatial planning is much more abstract and demands longer term commitments among the local governments. The decision to venture into spatial planning shows that the Joint Secretariat has been successful in building collaborative spirit among the three local governments, which may be a very important lead into even stronger roles for the Joint Secretariat.

The ultimate role of the Joint Secretariat would be the establishment of intergovernmental agencies which are fully responsible for specific infrastructure developments and services delivery, such as public transportation, water and sewerage, solid waste disposal, etc. Given the inward looking orientation of the local governments, the goal of establishing specific intergovernmental agencies is indeed a big challenge. Local governments need to begin looking at their neighbors more as partners rather than competitors. In the Yogyakarta urban region the executive branches of the local governments seem to have already caught the spirit of partnership. However, the local parliaments are more difficult to be steered toward more open collaboration toward neighboring local governments.

3.8 Conclusions: Sustainability in Question

It is clear from the above discussions that the goal of sustainable urban regeneration for the Yogyakarta urban region is not easy to achieve. A number of difficult challenges confront the local governments and the communities. First, there is need to revive the tourism related economy. Innovations can enhance tourist attractions by providing more variety and depth to the various cultural spaces and traditions, blending with modern lifestyles and convenience of access and information, and catering to international as well as domestic visitors. Handicraft industries related to tourism need to be encouraged and enabled to access wider markets including international markets, through wider networking for design, quality, and good production processes. Associations of handicraft producers should play an important role in creating effective networks among producers, designers, and marketers.

Second, the other important economic sector is higher education. As the decline in the number of student applicants is mostly due to the success of the family planning program in the 1980s, the colleges and universities in Yogyakarta need to shift from dependence on the number of students to prioritizing educational quality and research. The density of highly educated people in Yogyakarta is likely to be one of the highest in the country, therefore the creation of centers for research and development would be facilitated by sufficient human resources and such centers would attract funding and additional highly qualified personnel.

Third, the historical background of the Province of Yogyakarta not only gives it a special status within the governmental system of Indonesia, but also gives the Province better ability to reach and influence policy at the local government level, unlike other provinces in the country. The philosophy of unity between subject and ruler makes the Sultan the spiritual as well political leader of the entire Province (formerly the kingdom of Mataram Yogyakarta) in the eyes of the population. The Province is thus able to facilitate horizontal relations among local governments, even though by law the structural power of the provinces towards the local governments within their jurisdictions has been much reduced. Physically, the long history of the city gave it a number of interesting buildings, spaces, a tradition of arts and handicraft skills, as well as a distinct way of life. This gave the Yogyakarta urban region a specific character that made it an interesting destination for domestic as well as international visitors. The preservation of these historical artifacts, skills, and way of life is therefore important as heritage and as an economic resource.

Fourth, the duality of the people of Yogyakarta gives an interesting character to the city and its surroundings. Manifest duality can be seen in the mixture of structured urban forms and structures with informal and unstructured traditional settlements within the city. Modern urban lifestyles co-mingle with traditional practices and behaviors. Informal sector economic activities are performed on public spaces against the existing but un-enforced regulations. Such urban landscapes can be seen in many dualistic cities in developing countries such as Bangkok, Mexico City, Jakarta and other cities in Indonesia. On the other hand, duality can also be seen as the cause of non-compliant behavior of people with respect to formal law and regulations. This tendency is encouraged by the slack enforcement of the law, especially those dealing with behavior in public such as traffic laws, building codes, etcetera. This can explain, as described above, why housing development was uncontrollable despite all the plans and regulations that have been applied.

Fifth, for the last thirty years or so spatial development in the Yogyakarta urban region has occurred in an uncontrolled and opportunistic fashion. Housing developments were strewn over the rice fields, some of which were actually productive irrigated fields. Methods of control used by the local governments struggle and fail to contain the mobile small informal developers who seek out small plots and built houses. The conversion of agricultural lands into housing properties reduce the agricultural productivity of the area and push farmers away from their lifelong profession into unemployment. Population distribution became dispersed over wide but un-served areas, making it difficult to provide sufficient services to the population. Besides the environmental hazard of groundwater contamination, the haphazard developments also increase runoff that passed through urbanized areas toward the south, overwhelming drainage systems. Although the Yogyakarta urban region is relatively free of floods because of its generally flat and gently sloping topography, uncontained increased runoff would course through streets and cause damage. All these problems emanate out of the housing development process, which in the end caters to a housing market that treats houses as commodities for investment rather than for the fulfillment of need for shelter.

Sixth, the local governments of the Yogyakarta urban region have struggled with the effort to control urban development, but the pressure of the market is so strong that local government is unable to establish effective control over the entire range of development actors. This may well be attributed to the nature of duality of the culture. When regulations are made and not sufficient capacities and resources are created to uphold the regulations, and when plans are drawn up but the implementation of plans is not consequently followed, a dualistic approach to public management is evident. When the written formal laws are often not sufficiently enforced by the government and non-compliers go free, that is the manifestation of duality between the written law and the convenient fulfillment of daily necessities. On the other hand, when the provincial government, which by law has limited powers of intervention upon the local government, is able to exert political pressure on the local government because of its spiritual influence, that is another manifestation of duality. That the three local governments of the city of Yogyakarta and the regencies of Sleman and Bantul, under the influence of the Provincial government, were able to agree to support the Joint Secretariat which will intervene in their development policies for the common good, is another example of duality, which set Yogyakarta urban region apart from other urban regions in the country.

Seventh, in the final analysis, the characteristic duality of the people of Yogyakarta, which is manifested in the physical characteristics of the city and in the form of spatial development of the region as well as in the public behavior of the people and the political and bureaucratic behavior of the government, is one of the most important factors that explain the existing form of urban development in the Yogyakarta urban region. On the other hand, the institutional innovation for managing urban development in the form of the Joint Secretariat as a first step toward integrated management of the urban region, is more or less a result of the duality between formal government institutions and the informal influence of historical precedents. The sustainability of Yogyakarta urban region as a center of tradition and culture as well as higher education and scientific research and development basically rests on the ability of the business, industrial, tourism, and educational communities together with the government to overcome the dire effects of dualism and to capitalize on the beneficial effects of dualism.

Notes

[1] Student housing in Indonesian universities are generally off-campus, even in the larger national universities.

[2] Studies have been done on the effectiveness of the development control measures, especially the development permit system in a number of local governments. See Arnawati (2003), Syarifuddin (2000), Srianto (2000), Suryati (2003), and Wahyudi (2002).

[3] Ministrial Decree no 4/1980 and its follow up bylaws created a huge demand for spatial planners. Such demand sparked the growth planning schools, especially in the vicinity of the planning school at the Bandung Institute of Technology, Bandung, West Java.

[4] Basically the deadlock was caused by differing approaches between the two ministries. The Ministry of Home Affairs was the supervisor of local governments, which will be the implementers of the law. The Ministry of Public Works was traditionally the carrier of spatial planning functions and expertise (in the framework of infrastructure development).

References

Arnawati, E. (2003) *Attitudes and Behavior of Home-owners on the Building Permit System in the city of Yogyakarta* (Unpublished master's thesis)

Java Reconstruction Fund (2007) *Rehabilitating Livelihoods in Post-Earthquake Yogyakarta and Central Java Provinces* (Report)

Jhingan, M.L. (1975) *The Economics of Development and Planning* (Indonesian translation by Guritno, D.). Raja Grafindo Persada

Mumford, L. (1961) *The City in History: Its Origins, Its Transformations, and Its Prospects*. Jovanovich

Rapoport, A. (1977) *Human Aspects of Urban Form: Towards a Man-Environment Approach to Urban Form and Design*. Pergamon Press

Shima, N. (2007) "Challenges and Scopes toward the Decentralized Regional Development Framework in Southeast Asian Countries: Several Lessons from Yogyakarta Special Province, Indonesia", in Enlil, Z.M. and Vaggione, P. (eds) *Cities between Integration and disintegration*. ISOCARP

Simon, H.A. (1981) *The Sciences of the Artificial*. The MIT Press

Sriyanto, Y. (2000) *Land Use Controls through Development Permit in the Town of Ungaran, Central Java* (Unpublished master's thesis)

Subanu, L. (2004) *Urban Dualism in Asian Cities: the Case of Yogyakarta* (Paper presented at the Workshop on Urban Regeneration in Asian Cities, cSUR, the University of Tokyo)

Suryati, T. (2003) *The Effectiveness of Building Permit System as an Instrument for Land Use Control in Tangerang Regency, West Java* (Unpublished master's thesis)

Syarifuddin (2000) *The Effectiveness of Building Permit System as Instrument for Spatial Arrangement in the Town of Bangkinang, Riau Province* (Unpublished master's thesis)

Urban and Regional Planning Program, Gadjah Mada University (2007) *Final Report: Spatial Development Controls on the Periphery of Yogyakarta* (Study sponsored by GTZ-Good Local Governance Program)

Wahyudi, T. (2002) *Evaluation of the Effectiveness of Regulation on Land Conversion Permit in Sleman Regency, Yogyakarta Special Province* (Unpublished master's thesis)

4. Problems and Prospects for Sustainable Regeneration of Traditional Markets in Yogyakarta City

Kawik Sugiana

4.1 Background: The City of Yogyakarta

The city of Yogyakarta is the capital of the Special Province of Yogyakarta. Located in the middle-south of the island of Java (Fig. 4-1). The number of population of the Yogyakarta city was 510.914 in the year 2002. Table 4-1 depicts the number of population of regional cities and medium sized cities in Central Java and Yogyakarta provinces in 1999. Yogyakarta city is surrounded by several neighboring districts, i.e. Sleman, Bantul, Kulon Progo and Gunung Kidul (Fig. 4-1). The administrative-territory of Yogyakarta city covers a geographic area of 32.5 km square. The number of population living in the city and its suburbs was about 700 thousands in 1998 (Chamber of Commerce DIY (1998)). The annual increase rate of population in Yogyakarta was 7.6 percent. Strong migration into urban areas, especially to the capital of Yogyakarta, is the result of

Table 4-1. Number of population of regional cities and medium sized cities in Central Java and Special Province of Yogyakarta (1999) (Source: Central Statistical Bureau (2000))

No	Cities	Area (Km^2)	Number Population	ofPopulation Density per km^2
1	Kota Semarang (RC)	373.67	1,455,994	3,896.47
2	Kota Yogyakarta (RC)	32.50	490,335	14,555.10
3	Kota Surakarta (RC)	44.03	488,168	11,087.17
4	Kota Salatiga (MC)	52.96	163,079	3,079.29
5	Kota Pekalongan (MC)	44.96	285,829	5,912.57
6	Kota Tegal (MC)	34.49	238,059	6,902.26
7	Kota Magelang (MC)	18.12	116,498	6,429.25

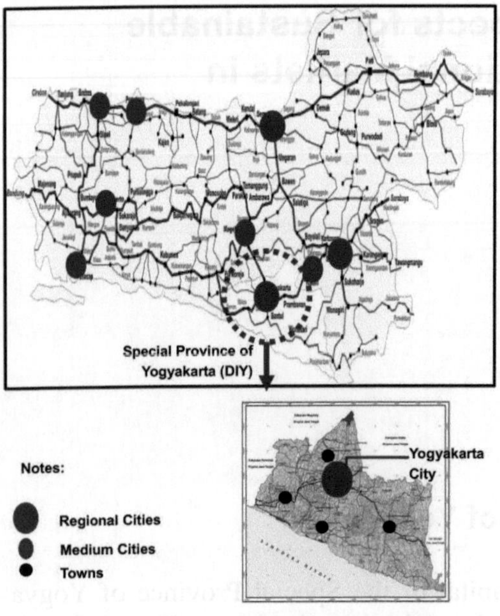

Fig. 4-1. Central Java and Special Province of Yogyakarta (DIY) (Source: Yogyakarta Special Province Development Planning Board (2004))

low living standard of the rural households.

Yogyakarta's regional economic development prior to the economic crisis in 1997 was marked by a distinct increased trend. The growth rates have amounted to 7 percent since the beginning of the 90's, which was above the national growth rate. The leading sectors of the economy are tourism, higher-education, and small and medium manufacturing. It should be stressed here that the informal sector of the economy and small and medium enterprises, like other urban areas in the developing countries, play important role in the urban economy, particularly in the period after the economic crisis.

Yogyakarta's current culture has its roots in the old Majapahit reign and in the Mataram kingdom. In Yogyakarta, Javanese culture and tradition is still alive and plays an important role in every day life. The cultural inheritances are found in the expression of artifacts and spatial arrangements, customs, arts, dances, believes and local institutions. A traditional market, a sultan palace, a mosque and a square are inherent parts of the central city of Yogyakarta.

Traditional markets in Yogyakarta urban area represent duality between the informal and the formal activities, as well as the traditional and the modern sectors. Indeed, the informal sectors tend to dominate trade activities in those markets. Based on their level of services, the traditional markets in Yogyakarta can be classified into three levels, i.e.: regional level, sub-district level, neighborhood level. The majority of them sell a wide range of product and services. However, more specialized markets are also found, dealing in song birds and animals, agricultural products, for example, and also in more expensive items. During the last decade, new shopping centers and super markets have been constructed and operated in central city and urban fringe of the Yogyakarta city. Although the price is relatively expensive, high and medium class families prefer to buy foods and goods in the supper markets, since they may get benefits in terms of better quality of goods and physical environment.

4.2 Traditional Markets in Yogyakarta City

4.2.1 Term and Definition of Traditional Markets

My approach to understanding the traditional commodity markets of Yogyakarta is based on Alexander's conceptualization (Alexander 1998) of the Javanese traditional markets as three analytically distinct systems: (1) trade, (2) traders and (3) trading.

The "trade" conceptualization treats the market as a system of material exchanges, examining the geographical distribution of marketplaces and the production and circulation of commodities. The main features of the traditional commodity markets of Yogyakarta are the wide variety of unstandardized commodities, the great range in the value of individual transactions, the large number of selling points, and the enormous numbers of traders, three-quarters of them women.

The perspective of "traders" highlights the social aspects of the marketing system: types of traders, and their relationships that link them in to complex trading networks. The Javanese traditional markets distinguish two main types of traders, *juragan* and *bakul*. *Juragans* are wholesalers. *Bakul* are predominantly women, dealing with varying success with a wide range of agricultural and manufactured commodities. It is unusual for *bakul* to cooperate to run a single enterprise; even when both spouses are *bakul* trading in the same commodities, they usually operate independent business. However, *bakul* often cooperate for such tasks as arranging transport, combine to purchase a large quantity of productions, or lend particular items to another. Many types of transactions take place between established trading partners. These relationships combine several levels of traders or producers into a complex financial system in which the debtor at one level becomes the creditor of other, smaller traders. The system serves a critical economic function of cheaply and efficiently distributing scarce finance through the marketing and production system as a whole. There are two main forms of institutional trading relationships: *Langganan tetap* (regular customers) and *ngalap-nyaur*. In a strict sense, the *ngalap-nyaur* relationships, characteristic of much petty trade, involves the provision of very short term credit. Goods are distributed in the morning, and paid for about noon after the market closes. All or some of the goods may be returned in lieu of payment. The more important *langganan tetap* relationship is a set line of interest-free credit. *Langganan tetap* is essential to maintain a high turnover, and traders unable to establish such relationships must work far harder and far longer for far a lower income.

"Trading", conceptualizes the market both as a structured flow of information and as a contested, culturally grounded system of meanings. As in the bazaar economies, accurate information, particularly price information, is the scarcest commodity in Javanese marketplaces. Because commodities are not standardized and are variable in supply, because prices are seldom marked on the product or stall, and because bargaining is the usual means of negotiating a deal, prices vary considerably from transaction to transaction. This is not a form of market failure (Alexander (1998)). Javanese traders see this unequal distribution of accurate price of information as an opportunity to profit from their superior knowledge. The success of a bakul dealing in potentially commodities depends on two commercial skills: her ability to negotiate prices when buying and selling, and her ability to maintain the reputation of being able to pay on demand (Rigg (ed) (1996)).

4.2.2 Operation of Traditional Markets in Yogyakarta City

Preliminary survey, based on observation and interview of buyers and sellers in Demangan, Kranggan, Sentul traditional markets, Yogyakarta, reveals that those three systems occur in the market, although some variations are found. Buyers, they may include households' wives, students, shop-keepers and also *bakul* --- who buy some goods in this market and sell them in the neighborhood markets --- state that the traditional markets are very important for their households' economy. They can get goods and services in terms of quality and quantity which they need in relative low prices. According to them, bargaining the price of un-standardized goods is a natural process and they are willing to do so in order to get lower prices. In other words, the traditional markets function as a mechanism for keeping the living cost relatively low in urban areas. Besides such a function, interviewees see the traditional market in an urban area as a public space for strengthening their relationships with families, friends and neighborhoods. For traders, sellers, particularly the informal actors, view traditional markets as strong opportunities to get jobs and income.

4.2.3 Spatial Distribution and Growth of Traditional Markets

The spatial distribution of traditional markets in Yogyakarta city is shown in Fig. 4-2. The area of services of the neighborhood markets covers almost the built-up areas of Yogyakarta city. In order to achieve

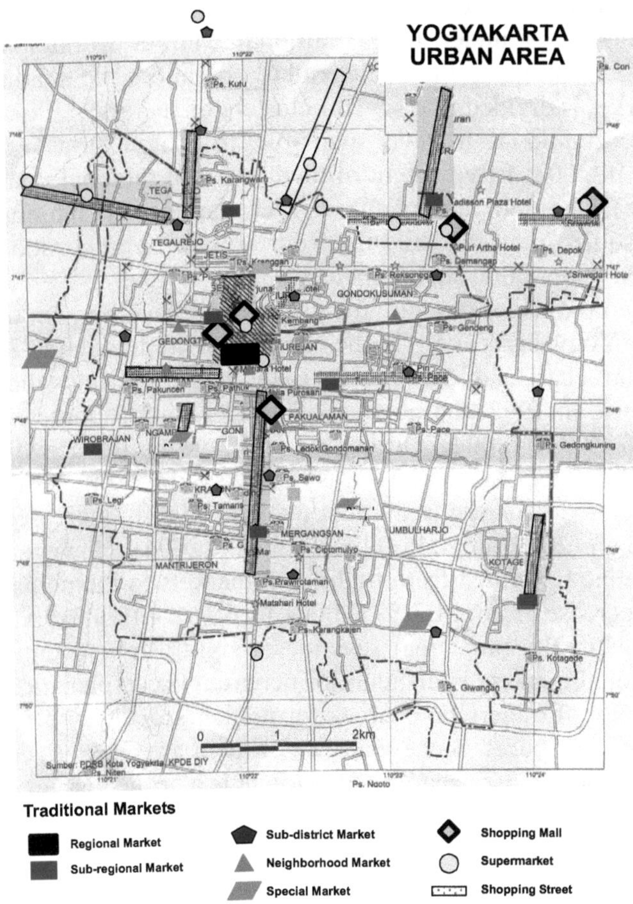

Fig. 4-2. Location of traditional markets and modern supermarkets in Yogyakarta (Source: Yogyakarta Special Province Development Planning Board)

Table 4-2. Revenues of the traditional markets (Rp.) (Source: Dinas Pasar Yogyakarta Munisipality)

Year	Traditional markets	Municipality	%
2000	3,643,705,788	24,790,130,000	14.70
2001	3,749,311,988	40,352,590,000	9.29
2002	3,922,250,949	56,377,000,000	6.96
2003	4,227,405,935	68,621,560,000	6.16
2004	4,444,708,067		

equitable access to the neighborhood markets, the Yogyakarta Urban Spatial Plan and the Development Guidelines stipulate that the neighborhood markets should be located in areas that they can effectively serve a certain number of people within 10 minutes walking distances. The management of the traditional markets is under the Traditional Market Municipal Agency (Dinas Pasar). To simplify the management operation, the spatial distribution of the market is divided into several clusters or sectors.

Total floor of the traditional markets available in Yogyakarta was 111,917 square meter in 2003; fifty percent of them was the regional market (Pasar Beringharjo). The floor area of sub-district markets was ranging from 1,100 to 5,900 square meters, while the floor area of neighborhood markets varies between 80 to 4,000 square meters.

Two indicators are employed to measure the growth and development of traditional markets in Yogyakarta, i.e. (1) the number of traders, and (2) revenues generated from charges of space usages. Table 4-2, indicates that after the economic crisis in 1997, the number of traders slightly increase from 12,748 (1998) to 13,242 (2004). During the last 7 years, the growth of total floor area for the traditional markets was relatively constant.

Total revenues generated from the traditional markets slightly increased from 3.64 billion Rupiah (2000) to, 4.45 billion Rupiah (2004). However, the percentage of revenue generated from traditional markets to municipal revenues tended to decrease from 14.7 % (2000) to 6.16 % (2003). Although its contribution to the municipal revenue is significant, revenue generated from traditional markets is difficult to increase. The explanation is, the willingness to pay of traders on charges for market spaces have direct impacts on their expenditures. Since majority of them are categorized as small and micro enterprises, user charges adjustment is a critical issue.

4.2.4 Market Chain

Market chain comprises two sub-system of interactions. First, interactions or linkages among actors in the rural areas --- they are primary producers and traders --- and actors in the urban areas --- i.e. traders in urban traditional markets. Second, interactions among various levels of traditional markets in urban areas.

In the case of Yogyakarta, most of agricultural products and goods come from outside of the Yogyakarta province. It is understandable, since the area of agricultural land surrounding the Yogyakarta city is very limited. Similarly, the number of large scale manufacturing industries is relatively small. Agricultural commodities including vegetables, fruits, rice available in the regional as well as district markets come from Central and East Java. Fish are originated from North Java Sea and south coastal area (Fig. 4-3). Textile, home appliances mostly come from Jakarta, Semarang and Surabaya.

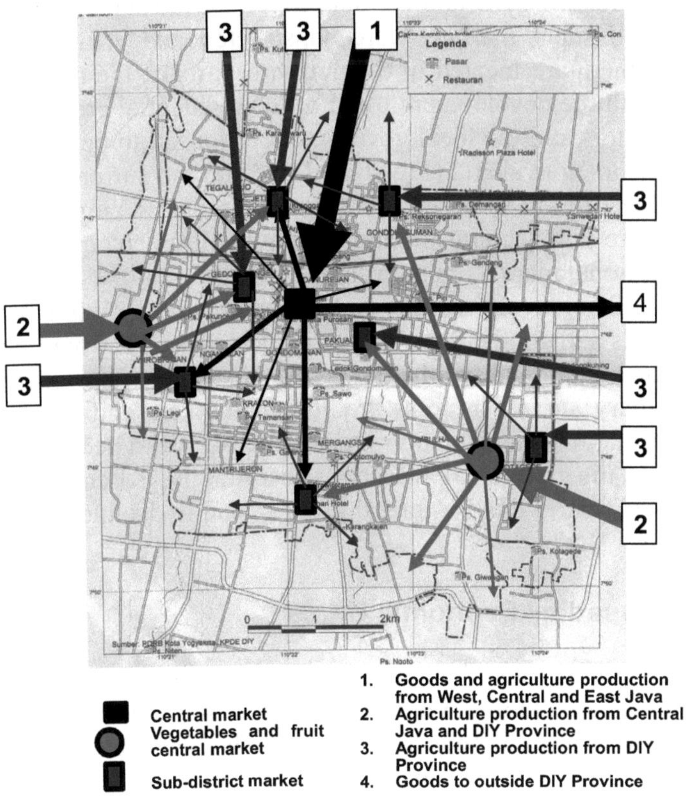

Fig. 4-3. Market chain in Yogyakarta City (Source: Dinas Pasar Yogyakarta Municipality)

4.3 Problems Faced by Traditional Markets

Current trend shows that several traditional markets in Yogyakarta face two major problems. First, problems associated with competition between traditional markets and shopping centers/super markets, and second, problems caused by economic crisis.

4.3.1 Development of New Super Markets and Shopping Centers

During the last decade, new shopping centers and super markets have been constructed and operated in central city and urban fringes of the Yogyakarta city. At the first stage, supermarkets and shopping centers were developed along the Malioboro CBD and the Sala street. At the sub-sequence phase, many mall and super markets were constructed and operated in other areas, most of them are located in the northern parts of the city and its suburbs, in line with the development trend of the residential areas. Recently, the private sectors and the local government proposed to build several other super-malls in Yogyakarta (Fig. 4-2). Based on the current trend, the increasing number of total floor area for super markets in Yogyakarta in the immediate future will account for approximately 400 percent, while the floor area of traditional markets remain stable.

Although the price is relatively expensive, high and medium class families prefer to buy foods and goods in the supper market, since they may get benefits in terms of better quality of goods and physical environment. Based on a preliminary survey in some traditional markets and limited secondary data-information, this situation has resulted in the decreasing number of buyers in the traditional markets. Accordingly, revenues received by traders, particularly those informal groups, tend to decline.

4.3.2 Problems Caused by Economic Crisis

Economic crisis strongly affected the number of job opportunities in large urban areas in Indonesia including Yogyakarta. Table 4-3 shows the impact of economic crisis on urban poor in Indonesia. The percentage of urban poor increased from 9.7 percent in 1997 to 21.9 percent in 1998. Informal job activities are considered as possible opportunities for many job seekers to satisfy their basic necessities. Strategic location of public spaces and traditional markets are chosen for their activities. Despite the decreasing number of buyers in the traditional bazaar markets, particularly

Table 4-3. Impact of economic crisis on urban poverty in Indonesia (Source: World Bank (2003))

Year	Percentage of population under the poverty line		
	Urban	Rural	Urban & Rural
1987	20.1	16.1	17.4
1990	16.8	14.3	15.1
1993	13.5	13.8	13.8
1996	9.7	12.3	11.3
1998	21.9	25.7	24.2

those medium and high income buyers, number of traders tended to increase.

This has resulted in several problems in the traditional markets: i.e. conflicts over the uses of spaces became more intent, revenues and the willingness to pay of traders on space usages tended to decrease. The ultimate result was the deterioration of physical-environmental quality of traditional markets.

4.4 Sustainable Regeneration of Traditional Markets in Yogyakarta

4.4.1 Why Regeneration of Traditional Markets?

Considering the uniqueness of traditional markets in Yogyakarta, the local government has initiated and implemented renovation programs to improve the condition of those markets. However, due to the complexity of problems encountered by traditional markets, the effectiveness of the programs is questionable. Conflict of interest among various actors including informal traders, small- scale retailers, gang members, street-vendors and pedicab drivers exacerbates the problems. Evidences show that physical improvement of the traditional market alone cannot solve the problems. Social approach is also required in this regard.

Falk (1998) states that "Regeneration is often treated as thought it is about property development, and if success can be measured by the level of investment, or even worst by the leverage ratio, the proportion of private to public investment. Cities constitute far more than just buildings in them, or the investment that these represent, it is naïve to measure and judge regeneration in physical term alone. Rather there is a need to think of cities in a more organic or holistic way, where success can be measured in the opportunities they give for the growth and fulfillment of human beings, and or both commercial and social enterprises".

Therefore, if there is no appropriate measure to revitalize the traditional markets, sustainable development for those markets as well as the Yogyakarta urban and regional economy may not be achieved.

4.4.2 Lessons from Previous Renovation Experiences

Indeed, local and provincial governments have initiated and implemented programs to renovate the traditional markets in Yogyakarta, including the Beringharjo regional market and some sub-districts and neighborhood markets. Beringharjo market was built in 1755-1792, in the period when the city of Yogyakarta was erected. It is a part of the original urban design of the Yogyakarta central city, and it is a unique icon of the Yogyakarta city and community. The renovation program including improvement of old buildings, construction of new buildings and relocation of traders have been initiated and implemented in responding to the need and development of the local government and society (Fig. 4-4). Successful results of the program as well as complaints by traders and community have been

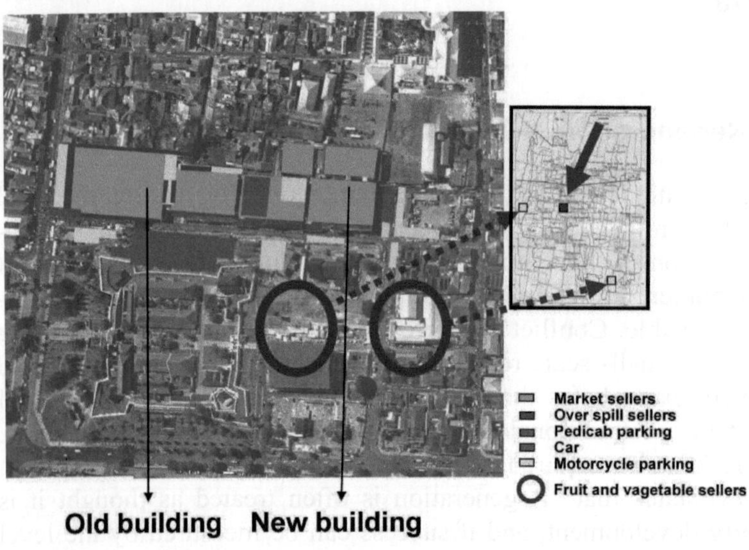

Fig. 4-4. Beringharjo was built in 1755-1792, during the period when the city of Yogyakarta was erected. It is a part of the 'first' urban design of Yogyakarta city. Renovation of the old building, construction of new buildings and relocation (of traders/sellers) programs have been initiated and implemented. Revitalization of Beringharjo traditional market

recorded.

The renovation program has brought about physical-environmental improvement of the buildings and their surroundings. Conflicts over the use of spaces were reduced, and over-spilled traders were relocated to other areas. Factors contributing the program success include: (1) intensive dissemination of the program; (2) effective interactions and negotiation between local governments and related traders; and (3) strong enforcement of regulations for controlling traders' activities.

Preliminary interview of selected traders and sellers also reveal several reasons regarding their objections to the renovation program: (1) they are afraid, after the renovation program they cannot afford to rent spaces previously used for their trade activities. Most of interviewees believe that the renovation program will increase the space rent. Without any commitment between those groups and the market manager as well as agreement among various stakeholders regarding the cost for renting spaces and the location for their activities, there will be no guarantee for them to get affordable spaces and strategic location; (2) moving to other location during the construction stage may result in discontinuity of relationships with buyers. Indeed, one precondition for maintaining a relationship between actors is a continued interaction; (3) moving to non-strategic location means loosing potential profits.

On the one side, increasing user charge of market space is likely opposed by traders; on the other side, revenue from space usage is significantly needed to improve the environment of traditional markets. Some questions emerge. How the trade off between those two opposing objectives suppose to be resolved? Is it possible for the local government to provide subsidy? Is it possible to invite the private sector and community to participate in the regeneration process?

4.5 Prospects for PPP Approach to Sustainable Regeneration of Traditional Markets in Yogyakarta

4.5.1 Rationale for PPP Approach

Financing gap in infrastructure and public facilities investment has been identified during the next five year national development plan (2005-2009). Financing Infrastructure Development Team (TPPI) at the national level indicates an investment requirement of Rp. 1,303 trillion over the next five year (2005-2009). Public investment through the state budget will provide only between Rp. 224,7 to Rp. 326 trillion. Between Rp. 179 to Rp. 332 trillion is expected to come from the domestic capital markets (banks, pension funds, insurance and mutual funds). Official development assistance (bilateral and multi lateral) will contribute Rp. 90 trillion. Thus, according to TPPI, there will be a financing gap ranging between Rp. 657 and Rp 887 trillion (50.4% -68.1% of the total budget) over the next five years. Such a gap, is expected to be fulfilled by the private sector contribution (KKPPI (2006)).

During the decentralization policy, the capacity of local governments to provide funding for local infrastructure and public facilities are relatively less than the national ones, especially the Yogyakarta Special Province and municipality. Therefore, contribution of private sectors' funding is essential for the provision, improvement, maintenance and operation of the public facilities including traditional markets.

Another reason for involving the private sectors in the regeneration process of traditional markets in Yogyakarta is the notion of efficient and effective management of public facilities. Empirical evidences in urban areas in developing countries showed that majority of public facilities operated and maintained by government entities tended to be inefficient and ineffective; while those provided and operated by private sectors seemed to be more effective and efficient.

4.5.2 Prospects

Given the limited availability of public fund, the national and local governments encourage the PPP approach to provision, improvement, maintenance and operation of public infrastructures and facilities. National and local policies related to improvement the business climate have been formulating, procedures and regulations with regard to participation of public sectors in infrastructures and public facilities development have been reformed during the last five years. The government of Indonesia issued Instruction of the President no 3/2006 to improve the general investment climate in Indonesia. The measures are: (1) to simplify the procedure to obtain permits and licenses; (2) to revise laws and regulations on investment and infrastructure provision; and (3) to provide tax incentives to promote direct investment in business activities in particular national priority areas (KKPPI 2006). Examples of policies to encourage goverments' PPP projects including: (1) the enactment of Presidential Regulation no 67/2005 which stipulates the participation of private sectors in public infrastructures provision, (2) formulation of guidelines to PPP projects; (3)

establishment of PPP center at the national level and PPP nodes at the sector and local levels. Opportunity to adopt PPP approach to public facilities provision, therefore, becomes more attractive to several provincial and local governments including the government of Yogyakarta.

Private sectors' funds, both foreign and domestic funds are available, including those coming from the Middle East countries. Recently, banks and private firms realize that investment in public infrastructures and facilities is economically and financially feasible if particular prerequisites are satisfied. The success of inviting investor, however, is dependent upon the ability and capacity of the local governments to convince investors regarding the feasibility of their PPP projects and to satisfy those prerequisites. Current evidences in several big cities show that many investors are willing to put their money in the development and operation of super markets and shopping malls. Some of them tend to recognize that investment in traditional markets is economically and financially feasible.

Although opportunities to achieve the expected objectives of sustainable regeneration of traditional markets through PPP approach are widely opened, several constraints are in existence. In order to attract the private sectors to contribute to the regeneration process of traditional markets in Yogyakarta, several preconditions related to better business climate should be satisfied. Based on opinions survey in several regions in Indonesia in 2002, "regional investment attractiveness" is determined by 5 factors, namely government institutions, socio-political condition, regional economic, labor and productivity, and physical infrastructure (Regional Autonomy Watch 2002). Factors government institutions and socio-political conditions are the most significant factors influencing the "regional investment attractiveness". Furthermore, this study elaborates 4 variables within those two factors which influence the level of attractiveness of a particular region. Those variables include: apparatus and services, local regulations, law certainty, and security. Low capacity and poor public services, conflicting among local regulations, law uncertainty and bad security may seriously decrease the level of attractiveness. The unfortunate conditions of the regions, imply that local regulations and local government bureaucracy should be reformed. During the last five years, efforts to reform the existing bureaucratic problems and constraints have already been initiated and implemented both at the national and local level, although the results may not be significant.

Another constraint is poor understanding among the local staff and community about Public and Private Partnership principles. Most of them believe that PPP projects will mainly serve the elite groups and disregard the needs of the local government and society. In PPP principles, there is an emphasis on partnership, of mutual benefit and gains that can be expected from a PPP project. In addition, risks should also be shared among parties involved in the contractual agreements. Unclear understanding by those parties about these principles, will result in difficulties to sustain the government's PPP program. Therefore, dissemination regarding the benefits and risks of having PPP projects should be conducted to various stakeholders.

Another essential issue of the PPP approach is controlling the behavior of private sectors. Indeed, private firms are profit making entities. Therefore, maximizing profit is their main objectives. Regulations and specific sanctions as well as a regulatory body are needed to direct and control the operation of private sectors in the PPP projects, including the delivery of quality and quantity of services, tariffs /charges setting and adjustment.

Competition among private sectors is essential in the bidding process of the PPP projects.. In the case of traditional markets, revenues generated from selling and leasing of market spaces are key factors in sustaining the regeneration process. In order to avoid opposition from traders, charges of space usages and the selling price of market space suppose to be kept affordable. Therefore, selecting qualified private providers should be conducted carefully in the project preparation stage. Through fair bidding, in the project preparation, qualified and efficient private provider could be chosen. Accordingly, the user charge or selling price of market space will be set at affordable prices. Fair biddings in the development projects, however, are usually constrained by various factors, including political pressures and groups' interests.

Previous experiences showed that many potential PPP projects failed to convince investors and funding institutions, due to poor designed of PPP document packages and project procurements (KKPPI 2006). In this regard, lack of ability and capacity of local government staff to provide sufficient document packages and procurements was the main constraint. This unfortunate situation implies that capacity building with regard to PPP approach should be provided to local government staff.

4.6 Conclusions and Implications

The traditional markets in Yogyakarta significantly contribute to the economic development and social-cultural meanings of the urban area of Yogyakarta as well as its surrounding regions. Current trend, however, shows that several traditional markets in Yogyakarta face two major issues. First, issue associated with competition between traditional markets and shopping centers/ super markets. Second, issue related to economic crisis which strongly affected the number of formal job opportunities in large urban areas in Indonesia including Yogyakarta. Despite the decreasing number of buyers in the traditional bazaar markets due to the competition with super markets, number of traders, especially those coming from the informal sector, tend to increase. This situation has resulted in several problems in the traditional markets: i.e. conflicts over the uses of space, total revenues generated from space rents decrease, and the ultimate result is the deterioration of physical - environmental quality of those markets.

Given the uniqueness and functions of traditional markets in Yogyakarta both as social-cultural instruments and economic mechanisms, efforts should be conducted to revitalize those markets. Indeed, local and provincial governments have initiated and implemented programs to renovate those markets. However, due to the complexity of problems encountered by traditional markets, several constraints are faced by the local and provincial governments. Central to the problem is the limited available public funds for sustainable regeneration of those markets.

The national and local governments are aware that a Public and Private Partnership (PPP) approach to public infrastructure and facilities provision is a potential alternative to solve the financing gap. Such an approach may also improve the efficiency and effectiveness of project deliverances and operations. In line with efforts of the national government to encourage public sector participation in the regional and local development, windows of opportunities are also widely opened for local governments including the Yogyakarta municipality, to adopt such an approach, particularly for sustainable regeneration of traditional markets.

Several constraints are identified including: (1) unattractive business climate in the region (i.e. apparatus and services, local regulations, law certainty, and security); (2) unclear understanding about the principles of PPP approach among local staff and community (i.e. equal partnership, mutual benefits, risks sharing, legal drafting); (3) inability of the local government to direct and control the behavior of private sectors to achieve the objectives of the PPP projects; (4) lack of capacity of local government staff to conduct fair bidding process in the selection of qualified bidders; and (5) lack of capacity to prepare and design sufficient document packages and procurements.

These unfortunate situations imply that the Yogyakarta provincial and local governments should be willing to conduct some measures to minimize constraints. During the decentralization policy after the reformation period, the provincial and local agencies are equipped with power and authorities to manage the development of their regions. Therefore, two main initial proactive actions should be initiated. First, to implement the good governance principles in order to encourage public sector participation in the regional development. Whilst, the governments should also simplify related regulations and streamline the bureaucratic procedures governing the participation of private sectors. The ideal institutional arrangements for public sector participation in the project provision is: policy making, regulator, contracting and operating roles should be performed by separate institutions. Second, to initiate and implement capacity building in terms of training and technical assistance related to the PPP approach to public infrastructure provision.

References

Alexander, J. (1998) "Women Traders in Javanesse Marketplaces: Etnicity, Gender, and Enterpreneurial Spirit" in Hefner, R.W. (ed) (1998) *Market Cultures*. Institute of Southeast Asian Studies

Falk, N. (1993) " Regeneration and Sustainable Development" in Berry, J et. al. (eds) *Urban Regeneration: Property, Investment and Development*. E&FN SPON

Chamber of Commerce and Industry, Special Province of Yogyakarta (1998) *Economic Potential Map of Yogyakarta*

KKPPI (2006) *Recent Developments in Indonesia's PSP Framework*

Regional Autonomy Watch (2002) *Daya Tarik Investasi Kabupaten/Kota di Indonesia: Persepsi Dunia Usaha* (in Indonesian)

Rigg, J. (ed) (1996) *Indonesian Heritage: The Human Environment*. Grolier International, Inc.

The World Bank (2003) *Kota-Kota Dalam Transisi: Tinjauan Sektor Perkotaan Pada Era Transisi di Indonesia* (in Indonesian)

The Asia Foundation (1999) *The Impact of Economic Crisis on Indonesian Small Medium Enterprises*

5. De Facto Urban Regeneration: A Case Study of Chiang Mai City, Thailand

Niramon Kulsrisombat

5.1 Overview of the Urban Regeneration Concept

Urban regeneration has been used as a means to address the decline of many urban areas. The problems include new social trends resulting from demographic change, the decentralization of people and jobs, and the move out of the city of younger and more able populations. A number of other problems include the continued physical deterioration of urban environments, the physical decay of towns and cities that results in a serious underutilization of scarce resources and creates pressure for the expansion of urban areas, and the decay or obsolescence of urban social and economic infrastructure. These urban problems are complex and interconnected by nature, and thus cannot be solved by a single-sector or single-agency approach. According to Roberts (2000), urban regeneration is defined as "*comprehensive and integrated vision and action which leads to the resolution of urban problems and which seeks to bring about a lasting improvement in the economic, physical and environmental condition of an area that has been subject to change.*" This implies an overall strategic framework for city-wide development and a departure from previous methods such as urban renewal, urban rehabilitation, or urban redevelopment which used short-term, fragmented, and project-based development. Significant elements of urban regeneration are described below.

First, urban regeneration employs a strategic approach. The need for a strategic approach for urban regeneration comes from the concerns that ad hoc projects without position within a broader vision of the region cannot effectively tackle urban problems. In other words, urban intervention which covers too small an area and too restricted a list of subjects cannot have a real impact on the problems of urban degeneration. This requires urban regeneration policy and action within a broader context as well as a comprehensive approach that deals with economic, environmental and social issues. The establishment of an integrated strategic initiative for urban regeneration at the regional level would appear to offer the prospect of better relating individual regeneration schemes within a wider regional context. It also provides a basis for bringing together all the necessary aspects of policy and implementation in a single portfolio. Therefore, urban regeneration will make more effective contributions to both overall regional regeneration and to the comprehensive treatment of individual problems areas (Roberts (1997b)). It is also equally important to establish mechanisms to ensure consistent urban regeneration and broad strategies. The devices include spatial planning, new organizational structures, and legal instruments.

Second, urban regeneration involves good local governance, which means to make decisions on the allocation of public resources based on the participation of various stakeholders including the different levels of government, the private sector, and civic sector. (Kidokoro (2005)). The multi-dimensional and complex nature of urban problems requires integrated, coordinated, and multi-faceted strategies involving multiple actors to address a wide range of social, economic, and environmental issues. Urban regeneration should integrate, vertically and horizontally, the policies, activities and resources of those partners into a comprehensive strategy. Closer involvement between the public and the private sectors, together with direct participation of local communities and the ability to cut across traditional policy boundaries, are all essential elements of an effective urban regeneration strategy (Bailey (1995)). Successful partnerships consist of clear allocation of responsibility and the involvement of local residents and community organizations as equal partners to ensure the appropriate goals and benefits from urban regeneration projects (Carter (2000)).

5.2 Urban Regeneration in Thai Context

The urban regeneration concept has been well developed in European, American, and some developed Asian countries such as Japan (Roberts and Sykes (2000)); (Kidokoro (2005)); (Sasaki); (Shimizu). In Thailand, however, the concept has been only recently introduced. According to Pokapanichwong (2002), urban regeneration has been discussed in Bangkok since the late 1980s when urban problems such as excessive urban sprawl, degrading quality of life, and social and economic decline among traditional communities in old urban areas were widely recognized. Several inner city improvement projects such as the Seminar on City Aesthetic and Humanizing Bangkok[1] were implemented in an attempt to halt the decline of old neighborhoods in inner city areas. Both focused on improving the quality of life in these old neighborhoods by providing public space such as pedestrian friendly streets, small parks, and civic plazas; reviving the neighborhood culture; and enhancing community capacity to enable participation in these urban revitalization projects. This trend was expanded into other large cities[2].

Two successful examples of inner city improvement projects should be noted here. Both of them are located in the Old Bangkok, Ratanakosin area. The first one is the revitalization of the Tha Tien community in Bangkok. It was initiated in 1997 and carried out through the collaboration between an academic institute and community organizations. The project employed physical environmental improvement and adaptive reuse of buildings and community building through awareness-raising (Pimonsathien). Another example is the revitalization of the Praeng Nara-Puthorn-Sappasarth neighborhood. The project was started in 1998 and carried out by community organizations in collaboration with academia and the city government. The main activities involved the conservation of traditional buildings and old shophouses; improvement of the physical environment; provision of infrastructure and open spaces; revival of neighborhood traditions; as well as community building through awareness-raising (Satapitanont et al. (2003). Although these projects have been implemented and yielded some good results, none of them has strategically addressed the issue by integrating city-wide development strategies. These projects were limited to project-based revitalization programs.

In contrast to the above examples, Chiang Mai city offers an interesting example of how the urban regeneration concept was employed to address urban problems. In 2005, the Chiang Mai city government launched a major urban regeneration project, called Chiang Mai High Point (CMHP). This project was developed out of the overarching provincial cluster development strategies. The CMHP regeneration project aims to induce economic and social development as well as promote the competitiveness of communities at a city-wide level through tourism development. It emphasizes policy formulation and implementation through a partnership-based approach among public, private, and academic institutes. It is, therefore, interesting to observe "*how the urban regeneration strategy has been formulated and implemented at the city level, and how partnership among various sectors has evolved through the formulation and implementation of urban regeneration projects.*" These issues will be investigated and discussed throughout the following sections.

5.3 Chiang Mai City and Urban Regeneration Issues

Chiang Mai city is the capital of Chiang Mai Province, thc second largest province in Thailand[3] (Fig. 5-1). It had been a capital of the Lanna Kingdom[4] since 1296, before becoming a province of Thailand in 1933. For this reason, the city is the largest and most culturally significant city in northern Thailand. It is located among some of the highest mountains in Thailand and stands on the Ping River, a major tributary of the Chao Phraya River. Thanks to the rich cultural history and natural resources, Chiang Mai city is the tourism center in northern Thailand (Fig. 5-2). Within the city, there is the old city with an area of 3.6 sq.km. which

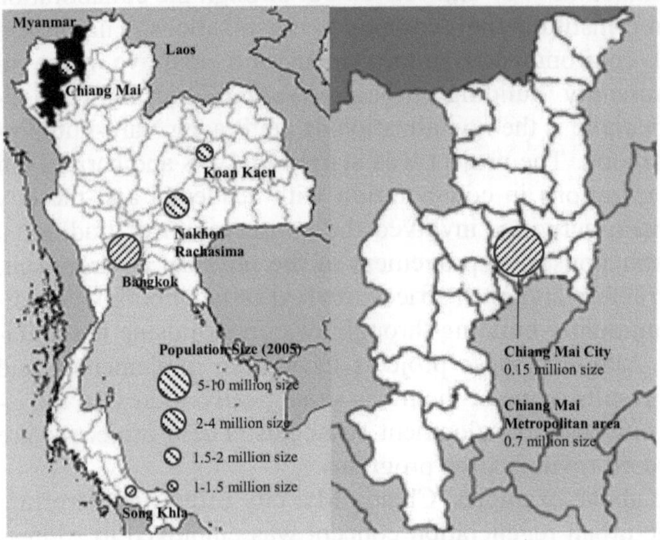

Fig. 5-1. Location of Chiang Mai Province and Chiang Mai City (Modified from Chiang Mai City Government Official Website (http://www.cmcity.go.th))

Fig. 5-2. Rich culture heritage in Chiang Mai (Courtesy of Korakoch Navanugraha

contains a number of Buddhist temples with traditional architecture[5]. Local markets and communities scatter around the old city producing and selling traditional handicrafts. Famous local products include silverware, lacquerware, woodcarving products, mulberry parasols, and silk and hand-woven textiles. Outside the city are natural attractions such as national parks, forest areas, waterfalls, and highland people of many ethnic groups. Tourist infrastructure and facilities are concentrated in Chiang Mai. For these reasons, Chiang Mai Province has the second highest tourist number after Bangkok. In 2004, Chiang Mai attracted 3.6 million tourists, an increase of15.2 per cent from 2003, which was equivalent to 31.3 per cent of the total tourist visits to Thailand (Tourism Authority of Thailand (2005)).

Because of its strategic location in terms the economy and security, Chiang Mai has been a target of several mega-projects from the Thai government. Since the beginning of the 1st National Economic and Social Development Plan in the 1960s, Chiang Mai has been designated as one of the regional growth centers to alleviate congestion in Bangkok as well as decentralize economic and industrial growth to the northern region. Infrastructure and public facilities were constructed to support economic development and industrialization, including airport, roads, dams, electricity generating plants, hospitals, and academic institutions. Since the late 1970s, Chiang Mai city has grown rapidly as it was designated as a *growth pole* for the northern region. In the mid 1980s, from the National Economic and Social Development Board (NESDB) initiative, the proposal to establish the twin city of Chiang Mai-Lamphoon was developed, followed by the Economic Quadrangle initiatives to integrate the economies of South West China, Myanmar, Laos, and Northern Thailand. In the same plan, Chiang Mai was also designated as a world tourist city full of art and culture. In the 8th Plan (1997-2001), Chiang Mai was designated as the hub for the Great Mekong Sub-region (GMS) Economic Cooperation Program and the South-East Asian Games (SEA Games).

After four decades, central government policies have helped induce some growth in Chiang Mai city, but at the same time have induced new challenges of sustainability in the city and surrounding rural areas. Following the Bangkok model, Chiang Mai city has become a primary city in the northern region where all economic activities are concentrated, while other areas remain economically under-developed. Fig. 5-3 shows a slight increase in business activity for every district of Chiang Mai Province; however, a large gap between Chiang Mai city and other districts still remains. In 2004, the total number of business activities in Chiang Mai city was 8.4 times larger than San Kamphaeng district, which is the second most economically developed district of the province. Consequently, Chiang Mai city attracts migration from the surrounding areas and other provinces[6].

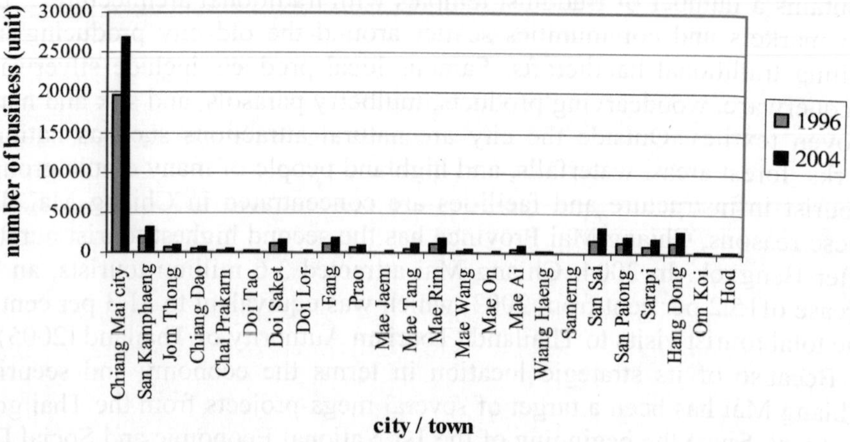

Fig. 5-3. The concentration of economic activities in Chiang Mai Province, 1996-2004 (Source: Annual Statistic Report of Chiang Mai Province 1996-2004)

Fig. 5-4. Deterioration of cultural heritage within inner city areas (left) and the severe flood in 2006 in Chiang Mai which partly resulted from great reduction of green space and forest areas (right) (Courtesy of Wongkanchena Kanchanaprachote)

The city expanded rapidly with new development areas, while traditional communities within the city core faced congestion and economic decline. Historical landscapes became degraded by unregulated construction. Air and water pollution became a serious problem due to urban congestion and lack of environmental management (Fig. 5-4). In March 2004, National Geographic's *Traveler* magazine named Chiang Mai as a city that had lost its charm and is "Getting Ugly."[7] This provoked a movement, especially among private and civic sectors, for regenerating the city[8].

5.4 From Regional Development Policy to De Facto Urban Regeneration

From a policy survey, it was found that the urban regeneration concept had not yet been formally included in Chiang Mai city's development policy. However, the concept was incorporated in the 9th National Economic and Social Development Plan (2002-2006), which aimed to strengthen regional competitiveness through the formation of provincial clusters to function as a key mechanism in increasing productivity, distribution of growth, and holistic management of resources at a regional scale. The National Economic and Social Development Board (NESDB)[9] initiated 19 provincial clusters based on pragmatic mix, economic and natural-based criteria. Regional development strategies were formulated in line with the National Development Plan in order to encourage participation among all social sectors: the public; private; and civic sectors; as well as academic institutions with NESDB regional branches acting as coordinators. The aims were to enable cooperation among provinces, focus on the genuine and full potential of local resources, and to underline needs and concerns of the local communities in order to bring about a more realistic development policy. Therefore, the regional development strategies were used as the overarching single framework for all provincial members to ensure the consistency of policy implementation.

The NESDB selected the Upper Northern Provincial Region, known as the Lanna cluster[10], for a provincial cluster pilot project because it had already been strategically placed in the GMS plan. It was the largest cluster, consisting of eight provinces: Chiang Mai, Chiang Rai, Lamphoon, Lampang, Mae Hong Son, Phrae, Nan, and Payao. Chiang Mai was designated as the center of the cluster. The new vision for the Lanna cluster was formulated under the motto: *Lanna Links Global Reach*, aiming to achieve sustainable increases in productivity and profitability at a regional scale. One of the major development strategies was cultural-based development by enhancing the value creation of tourism and handicrafts[11]. In order to translate the strategy into action plans, the Northern Economic and Social Development Office (NESDB regional branches) formed a working team of representatives from various regional sectors including: the Northern Department of Industrial Promotion; the Northern Handicrafts Manufacturers and Exporters Association (NOHMEX)[12], and the Faculty of Fine Arts, Chiang Mai University. All of them are experts in the fields of Lanna culture, tourism, and handicraft development.

The major scheme, Lanna High Point, is a holistic regional development scheme based on tourism and handicraft development. Inspired by the

High Point market in North Carolina (USA)[13], Lanna High Point aims to create live exhibitions of northern handicrafts where tourists can shop at real production sites in the traditional communities of Lanna province members. The regional development scheme consists of two major components: (1) regeneration of the economically declining traditional communities within the city; and (2) distribution of growth to outlying communities. Comprehensive tourism networks are to be created among traditional communities along with the development of tourist infrastructure, handicraft products as well as human resources. It is expected to bring broad-based economic growth to the whole region. Lanna High Point is divided into three phases covering the years 2005 to 2009. The first phase started with Chiang Mai city and its neighboring towns under the theme: Chiang Mai High Point (CMHP) (2005-2006).

In order to translate the regional strategy into action at the city level, a working team[14] held a series of meetings to consult with local governments of the target areas. Each local government was required to propose an artisan community as a pilot project under the CMHP. Projects came up at many communities located within and outside Chiang Mai city, as follows: (1) Wua Lai community (silversmith and lacquerware: Chiang Mai city); (2) Ban Ton Pao community (mulberry paper: Ban Ton Pao Tambon Administration Office, or TAO[15]); (3) Ban Pa Bong (woodcarving: Sarapi TAO); (4) Ban Kiw Lae Noi (Mudmee cotton cloth: Ban Kiw Lae Noi TAO); and (5) Ban Muang Koong (ceramics: Hang Dong TAO). In the end, all parties agreed on the CMHP regeneration scheme which aimed to: (1) increase competitiveness of Chiang Mai city and its neighboring towns as tourism and handicraft centers by improving its traditional communities, community products, as well as infrastructure and public facilities; (2) link all communities into a comprehensive tourism network (Chiang Mai city government (2005).

5.5 Urban Regeneration Implementation Mechanisms

In order to facilitate regional regeneration policy implementation, three mechanisms were established: (1) the CEO governors and the CEO budget; (2) the Development Strategy Driven Center; and (3) Provincial Comprehensive Plan.

First, the CEO governors and the CEO budget[16] were established as fiscal instruments. The CEO governorship shifted the role of the provincial governor from policy facilitator and coordinator into a more significant role as policy formulator, implementator, and administrator - in other words, as chief executive officer (CEO) of the province. The CEO governor acted as the secretariat of the Prime Minister. This scheme aimed to disperse authority to the local government, where decision-making was brought closer to the people. These frameworks would help facilitate effective coordination in development activities at the provincial and local levels, which would be effectively responsive to the public's needs. For the Strategic Provincial Cluster Budget, or the CEO budget, extra budget was allocated for implementing the projects which were in line with regional development strategies according to the provincial cluster development.

Second, the Development Strategy Driven Center (DSDC) was set up as a coordination mechanism for all developments at the supra-provincial level. Also, it was to strengthen the link between regional policies and project implementation. Since the office of provincial governor was always occupied by routine responsibilities, the DSDC aimed to help them to facilitate, follow up, monitor, and coordinate all sectors in project implementation (Fig. 5-5). The DSDC was established by the CEO budget. The DSDC's team consisted of professional staffs which were appointed by the Northern Economic and Social Development Office. It was established in every province within the Lanna cluster, with Chiang Mai as the headquarter.

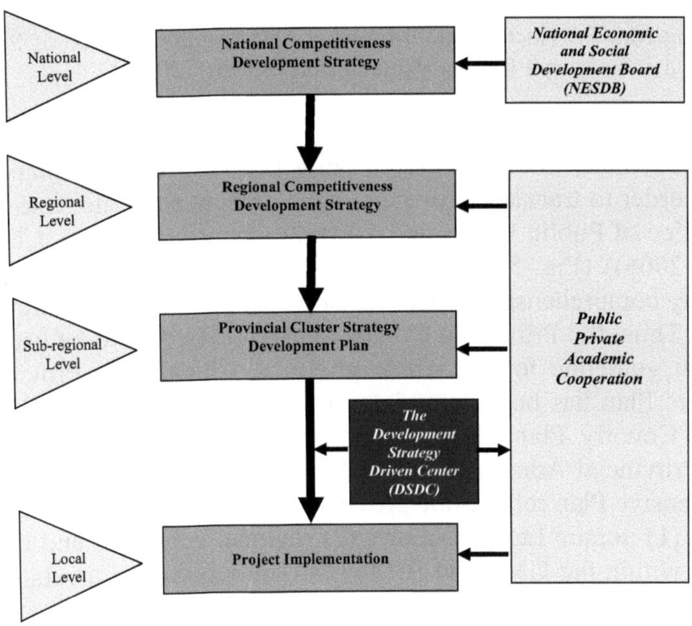

Fig. 5-5. Policy formulation and implementation process of the provincial cluster initiatives (Modified from the Northern Economic and Social Development Office (2003))

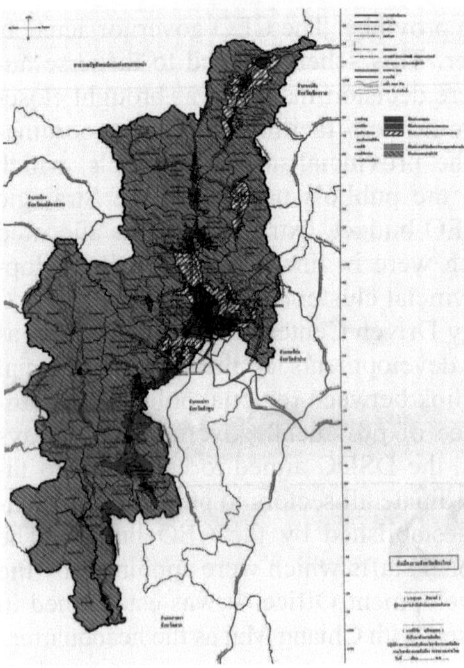

Fig. 5-6. Chiang Mai Provincial Comprehensive Plan (Source: Office of Public Works and Town & Country Planning (Chiang Mai Branch) (2006))

Third, the Provincial Comprehensive Plan was established as a planning instrument in order to translate provincial development strategies into spatial plans (Office of Public Works and Town & Country Planning, Chiang Mai Branch (2006)) (Fig. 5-6). Only small areas in each province were covered by city comprehensive plans, which would result in uncoordinated development. Thus, the Provincial Comprehensive Plan was to be used as a development guideline for the whole province. Chiang Mai Provincial Comprehensive Plan has been formulated by the Office of Public Works and Town & Country Planning (Chiang Mai Branch) together with the Chiang Mai Provincial Administrative Office. The concept of the Provincial Comprehensive Plan reflects the provincial policy, and consists of four main themes: (1) unique Lanna Culture; (2) regional economic and transportation hub within the GMS and BIMSTEC sub-region; (3) diversity of agricultural products; and (4) excellent tourism service and infrastructure. The Provincial Comprehensive Plan consists of: (1) conceptual plan which designates the location of urban communities, hierarchical order and linkage between them; and (2) land use plan[17].

5.6 Implementation of Urban Regeneration Projects

A survey[18] found that there are a number of projects relating to urban regeneration that have been implemented by various organizations, yet with different approaches. These projects can be divided into three categories according to the nature of plan formulation and implementation: (1) local government-led; (2) private sector-led; and (3) central government-led, as shown in Table 5-1.

5.6.1 Local Government-led Projects

Local government-led projects include the CMHP regeneration project with the five local governments as described in the previous section. According to the survey, only the Ban Wua Lai silversmith community project has been implemented by Chiang Mai city government. In order to tackle social and economic decline in the area, Chiang Mai city government launched two revitalization programs. First, Wua Lai Walking Street aimed to promote the community's silverware to the public by closing the main street for pedestrian use on Saturday nights. Chiang Mai city

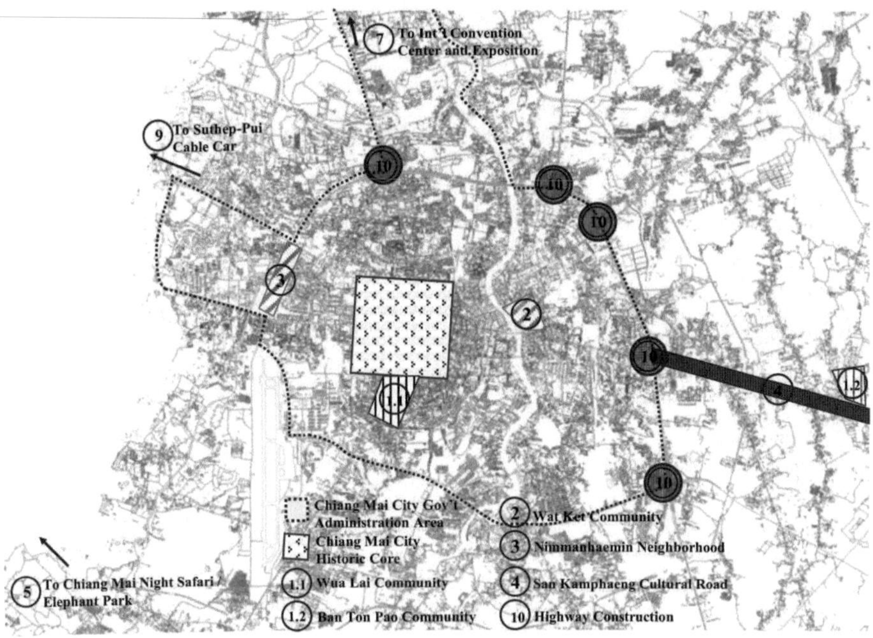

Fig. 5-7. Location of projects related to urban regeneration within Chiang Mai city and periphery

Table 5-1. List of projects related to urban regeneration within Chiang Mai city and neighboring areas (Source: Chiang Mai High Point Project Report (2006), Apawacharuth (2005), with author's survey in December, 2006)

Project	Type of Development	Initiator	Implementor	Fund	Status
1. Chiang Mai High Point					
1.1. Wua Lai community revitalization	Physical environment upgrade / Local product improvement / Tourism marketing / Tourist infrastructure improvement	[L] : Chiang Mai municipality	[L] : Chiang Mai municipality	CEO budget / related government including Tourism Authority of Thailand (regional branch) and the Northern Industry Promotion Department	Only Wua Lai community revitalization project has been undertaken, while other projects have not been implemented yet
1.2. Ban Ton Pao community revitalization		[L] : Ban Ton PaoTAO	[L] : Ban Ton Pao TAO		
1.3. Ban Pa Bong revitalization		[L] : Sarapi TAO	[L] : Sarapi TAO		
1.4. Ban Kiw Lae Noi revitalization		[L] : Ban Kiw Lae Noi TAO	[L] : Ban Kiw Lae Noi TAO		
1.5. Ban Muang Koong revitalization		[L] : Hang Dong TAO	[L] : Hang Dong TAO		
2. Wat Ket community revitalization	Urban conservation / Adaptive re-use of old houses and shophouses	[P] : Community-based organization / private entrepreneurs	[P] : Community-based organization / private entrepreneurs	Voluntary donation for urban conservation / Private investment	Underway
3. Nimmanhaemin neighborhood revitalization	Adaptive re-use of old houses and shophouse	[P] : private entrepreneur	[P] : Community-based organization / private entrepreneur	Private investment	Underway
4. San Kamphaeng Cultural Road	Road construction / Landscape improvement	[C] : Chiang Mai Provincial Office	[C] : Highway Authority (Chiang Mai Branch)	CEO Budget	Completed
5. Chiang Mai Night Safari	Large scale development	[C] : Ministry of National Resources and Environment	[C] : Designated Areas for Sustainable Tourism Administration (DASTA)	CEO Budget	Completed
6. Elephant Park	Large scale development	[C] : Tourism Authority of Thailand	[C] : Tourism Authority of Thailand	CEO Budget	Not implemented yet
7. International Convention Center and Exposition	Large scale development	[C] : Tourism Authority of Thailand	[C] : Tourism Authority of Thailand	CEO Budget	Underway
8. Northern SMEs Business Development and Distribution Center	Large scale development	[C] : Ministry of Industry	[C] : Office of Industry Promotion	CEO Budget	Not implemented yet
9. Suthep-Pui Cable Car	Large scale development	[C] : Tourism Authority of Thailand	[C] : Tourism Authority of Thailand	CEO Budget	Not implemented yet
10. Highway Construction Project	Highway constructions	[C] : National Highway Authority	[C] : Highway Authority (Chiang Mai Branch)	CEO Budget	Underway

Note: **[L]**, **[P]**, and **[C]** stand for local government-led, private sector-led, and central government-led project respectively.

Fig. 5-8. Wua Lai community after the implementation of landscape improvement programs

government provided support for public relations, garbage collection, and electricity. Second, landscape improvement provided a sculptural landmark as well as benches for tourists (Chiang Mai city government (2004)) (Fig. 5-8). The project, however, was not so successful in revitalizing economic activities. From interviews with local residents and entrepreneurs, these revitalizing programs did not tackle to the major problems including cultural heritage deterioration, through-traffic, and migration of young people to the San Kamphaeng area[20]. Wua Lai Walking Street helped induce a small increase in sales during the weekends, but weekdays remained quiet[21]. Also, some mentioned that the program overlaps with the Tha Phae Walking Street[22] which is held regularly every weekend and is located very close to the Ban Wua Lai community.

5.6.2 Private Sector-led Projects

According to the survey, there were two neighborhood revitalization projects which were not included in the CMHP regeneration project, yet yielded some remarkable outcomes. These projects include the revitalization of the Wat Ket community and Nimmanhaemin neighborhood. Both areas have been carried out exclusively by private entrepreneurs and community-based organizations. In the case of the Wat Ket community[23], the urban revitalization project was started by the Wat Ket Conservation Club, which was established by community residents in collaboration with the Lanna Architects Association and the Social Research Institute of Chiang Mai University in order to safeguard the cultural heritage of Wat Ket. The project consists of renovating Wat Ket's old pavilion and establishing the Wat Ket Museum. Consequently, old shophouses were renovated and kept in a good condition[24]. Furthermore, the Livable Community Project was

Fig. 5-9. Wat Ket community (Courtesy of Wittaya Duangthima)

conducted by community residents to research and disseminate community history and culture. These brought Wat Ket public recognition and turned previously derelict areas into investment spots. At present, several shophouses have been adaptively reused as home décor and handicraft shops, small restaurant and outdoor cafes, blending with other local shops and residences. The area became popular among young people and tourists (Fig. 5-9).

In the case of the Nimmanhaemin neighborhood[25], it was initially a quiet, low-density residential area. In 1995, small entrepreneurs started opening 4 home décor shops which expanded to 22 shops in 2006. Most of these famous home décor and handicraft shops were nationally recognized. The Nimmanhaemin neighborhood later became a new home décor and handicraft area of Chiang Mai. The shop-owners established the Committee of Nimmanhaemin to take care of public facilities as well as organize the Nimmanhaemin Art and Design Promenade (NAP)[26], a special annual fair that turns Nimmanhaemin street into a pedestrian street showcasing art and handicrafts products. It has provided a stage for talented designers, not only for those with houses, shops, and galleries in Nimmanhaemin street but also for groups of artists and artisans from around Chiang Mai selected by the Committee of Nimmanhaemin. The neighborhood has developed into a shopping district with great mixture of small-scale shops and activities, and is one of the most lively hang out spots in Chiang Mai city for town people and tourists.

5.6.3 Central Government-led Projects

Besides small-scale revitalization projects mentioned above, the survey identified a number of mega-projects which have been initiated and implemented by the central government. So far, only two projects have been completed: San Kamphaeng Cultural Road and Chiang Mai Night

Fig. 5-10. San Kamphaeng Cultural Road (left) and Chiang Mai Night Safari (right) (Source: Author's survey (left) and http://www.chiangmainightsafari.com/download (right))

Safari (Fig. 5-10). San Kamphaeng Cultural Road[27] covers a physical improvement area of 12 km. along the road. The project aimed to strengthen the links between the historic core of Chiang Mai city and the San Kamphaeng district, a new center for large-scale production of handicrafts. It was designed to encourage tourists to use a bicycle lane and wide pedestrian walkway. It was also expected to increase tourist visits in traditional artisan communities in the San Kamphaeng district. The project was initiated by the Chiang Mai Provincial Government; however, it was widely criticized for having a wasteful and expensive construction budget as well as lacking transparency[28]. Also, it contradicted the CMHP regeneration project which aims to revitalize the Wua Lai community, where the similar products are produced.

Chiang Mai Night Safari was a large-scale project with an area of about 1 sq.km in the Suthep-Pui National Park. The objectives of the project were to establish a new alternative for tourists during the night time and to be the world's standard in tourism conservation and research on animals and plants. Chiang Mai Night Safari was initiated in 2003 by the Ministry of National Resources and Environment, and the Office of Conservation Area Management. The project has been strongly opposed by several civic organizations due to several reasons. First of all, the project involved large-scale construction which was not compatible with the National Park area and inevitably harmed bio-diversity, and intensively consumed energy and water, as well as other resources. Secondly, the project contradicted the cultural capital image of Chiang Mai city. Lastly, the project lacked transparency because the responsible organizations held very limited public hearings among local people who lived on the construction site[29].

5.7 Discussion

Although the urban regeneration concept has not appeared in the formal Chiang Mai city development policy, Chiang Mai city government launched the CMHP within overarching provincial cluster development strategies which resulted in the implementation of a number of urban revitalization and development projects within Chiang Mai city and its neighboring towns. These projects have been carried out by various organizations; however, they were operated independently without linkages among them. As a result, these projects failed to address long-standing urban problems: inner areas deterioration and uneven growth. The problems of urban regeneration policy formulation and implementation in Chiang Mai city are described as follows.

First, there was a lack of strategic approach for regeneration policy formulation at the city level. Even though the CMHP regeneration project was developed out of an overarching regional regeneration policy, local government failed to address the relationship with other mainstream development policies such as land use, transportation, and cultural heritage conservation at the city level. Instead, it represented a series of piecemeal projects without a long-term vision of the project's impact on the communities and city-wide area. The superficial revitalization program of Ban Wua Lai exemplifies this problem. It is unrealistic that mere landscape improvements would resolve long-standing problems facing communities.

Second, there was a lack of effective instruments to ensure consistency of urban regeneration policy implementation at the project level (Fig. 5-11). Although the Provincial Comprehensive Plan was initiated to serve this purpose, its implementation was highly questionable for the following reasons. First, the Provincial Comprehensive Plan was too broad to effectively achieve project goals. Similar to the City Comprehensive Plan, it aimed to promote sound development through land use controls and permits. In fact, a Specific Plan would be more appropriate to bridge the gap. Second, the methodology was rather vague considering the employment of a similar land use zoning control from the Comprehensive Plan at the provincial level. A Structural Plan would be a more appropriate device. Third, plan formulators (DTP Chiang Mai Branch) and regulators (local government) were different organizations. The possibility for inconsistency in implementation due to misunderstanding and lack of commitment is high. The last problem deals with enforcement. The Town Planning Act of 1975 defines a Comprehensive Plan as "the outlines for urban-rural development in property benefit, transportation, public facilities, service and environmental status." Therefore, the plan implies enforcement at the city level,

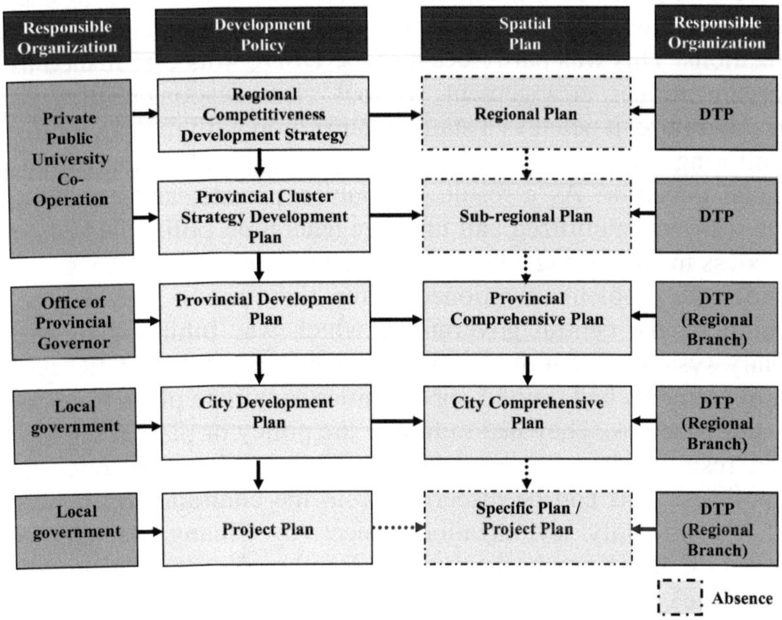

Fig. 5-11. Local governance inhibited in the policy formulation and implementation process

not the provincial level.

Third, there was insufficient local governance in formulating and implementing urban regeneration projects. This was reflected by a lack of vertical coordination in the process of policy formulation and implementation. Although the regional policy formulation has shifted toward an intensive collaboration among multiple, area-based organizations such as the Northern Economic and Social Development Office, NOHMEX, the Northern Department of Industrial Promotion, Office of Industrial Promotion and Chiang Mai University, the involvement from local government was very limited since they were invited at the very end of policy formulation process[30]. This resulted in less understanding and commitment from the local government in implementing the policy at the city level. This can be observed from the little amount of project completion by the local government.

Another limitation was the lack of horizontal coordination. There was little participation between the municipality, private organizations, civic and community-based organizations in the formulation of urban regeneration at the city level[31]. Also, there was little communication and coordination between Chiang Mai municipality and other local agencies whose communities were designated in the CMHP projects. This could be because of the inactive role played by the DSDC in mobilizing cooperation with related organizations. This was partly because the DSDC was established as an ad hoc organizational arrangement without clear responsibilities or legal power. Therefore, it acted as a static organization with mainly routine tasks of monitoring the plan and collecting data to submit to the office of the provincial governor. As a result, the public, private, and community resources were not mobilized and urban regeneration policy lacked comprehensiveness in addressing city-wide as well as community issues.

Besides the problems mentioned above, there was a high degree of intervention by the central government which was inhibited by the CEO budgetary system. When the budget was controlled by the CEO governor, local governments had limited opportunities to initiate plans based on their needs and concerns. They had to follow the policy or plan in the top-down manner, resulting in a project which may contradict with local development policies. This can be discerned from the contradiction between the Wua Lai community revitalization project (by Chiang Mai city government) and the San Kamphaeng Cultural Road project (by Chiang Mai provincial office), which aimed to promote similar products, or other megaprojects that would undermine the intention to limit urban sprawl.

All these problems prevented Chiang Mai city from effectively translating regional development policy into action and benefiting the intended recipients of individual regeneration schemes and the region as a whole. Two recommendations for future urban regeneration are suggested. First, a strategic framework should be established at the city-wide level in order to make "links-setting contexts to foster relationships; and strategic vision-mobilizing ideas about the future" (Healey et al. (1995)). The strategic framework will allow wide-range participation from various organizations and communities in city-wide policy formulation, thereby providing a bridge between top-down and bottom-up initiatives. Furthermore, implementing coordination devices should be provided. Two-tier spatial plans, consisting of city-wide and project levels, can ensure consistent policy implementation. Second, a city-wide coordination mechanism should be set up to promote better local governance. This could be the establishment of the Development Strategy Driven Center at a city-wide level adapted from the existing DSDC structure. It should consist of a wide range of representatives from all sectors (public, private, academic and volunteer organizations) and should be selected through an election. It will serve as a vertical and horizontal link and, therefore, can strengthen the links between local initiatives and regional policies.

5.8 Conclusion

Despite the absence of the urban regeneration concept from the formal Chiang Mai city development policy, Chiang Mai city government has launched a major urban regeneration project, the Chiang Mai High Point (CMHP), out of the overarching provincial cluster development strategies which aimed at reviving inner cities areas as well as distributing of growth to outlying communities. The study revealed that, however, the CMHP failed to achieve its goals due to three reasons: (1) a lack of a strategic approach for regeneration policy formulation at the city level; (2) a lack of effective instruments to ensure consistent urban regeneration policy implementation at the project level; and (3) insufficient local governance in formulating and implementing urban regeneration project as well as a high degree of intervention by the central government. If urban regeneration is to be continued in future and yield sustainable outcomes, two recommendations are made: (1) the creation of a strategic framework to bridge top-down and bottom-up initiatives and translate the city-wide policy into spatial plans; and (2) the formation of the Development Strategy Driven Center at a city-wide level as a policy coordination mechanism.

Notes

[1] Seminar on City Aesthetic City and Humanizing Bangkok projects were initiated by the Bangkok Forum in the Ban Krua, Talat Noi and Yaowarat communities (Pakapanichwong (2002)).

[2] The trend expanded to Songkhla, Chiang Mai, and Khon Kaen.

[3] Geographically, Chiang Mai Province is the second-largest province of Thailand. It is located in the north of the country. Neighboring provinces are Chiang Rai, Lampang, Lamphun, Tak and Mae Hong Son. It also shares borders with the Shan state of Myanmar. Chiang Mai Province is divided into 22 districts (*Amphoe*) and 2 minor districts (*King-Amphor*). The districts are further subdivided into 204 communes (*Tambon*) and 1,951 villages (*Muban*).

[4] Lanna Kingdom, or *the Kingdom of a million rice fields*, used to be one of the most powerful kingdoms in Southeast Asia. The kingdom covered Northern Thailand, North-Western Laos, eastern Burma, and southern Yunnan. It enjoyed its Golden Age throughout the 15th century. However, Lanna was caught between Burmeses and the Ayutthaya of Thailand during the 16-18 century. In 1774, King Taksin of Siam took Chiang Mai back from Burma and used it as a vassal town. In the reign of King Rama V (1868-1910), Chiang Mai's status was upgraded from a vassal town to a part of

Monthon Phayap. When the Thai government abolished this unit, Chiang Mai became a province of Thailand in 1933.

[5] There are 228 Buddhist temples located in Chiang Mai city.

[6] This Chiang Mai Metropolitan Area has a population of nearly 700,000, nearly half the population of Chiang Mai Province (Office of Public Works and Town & Country Planning (2006)).

[7] National Geographic's *Traveler* magazine (March 2004 edition) rated Chiang Mai and Phuket, two Thai world-famous tourist destinations, as "Getting Ugly." A scoreboard of destinations around the world was created by Leeds Metropolitan University in the UK with interviews of professionals on sustainable tourism and destination quality. (The Irrawady News Magazine Online Edition http://www.irrawaddy.org/aviewer.asp?a=3743&z=24)

[8] Interview with Wiwat Pongruengkiat, the vice president of Chamber of Commerce, Chiang Mai. The same issue was also mentioned in Apawacharuth 2005.

[9] National Economic and Social Development (NESDB) is supervised by the Office of the Prime Minister.

[10] All provinces share a strong cultural identity of *Lanna*.

[11] Four main strategies of the Lanna Cluster Development consisted of: (1) creating new economic base; (2) enhancing value creation of tourism and handicrafts; (3) strengthening the conventional economic activities; and (4) preserving and rehabilitating the natural resource base.

[12] The Northern Handicrafts Manufacturers and Exporters Association (NOHMEX) was established in 1991. The founding members consisted of 13 handicrafts producers in San Kamphaeng district, Chiang Mai. The origin of the establishment was to solve the production problems among member factories and create bargaining power between the association and buyers in provincial areas. In 1994, NOHMEX was upgraded to a registered association and expanded membership to exporters of handicrafts. NOHMEX reached more than 250 members (Interview with Napong Snguannapaporn, the first president of NOHMEX).

[13] High Point Market is the largest and most comprehensive exhibition of residential home furnishings in the world. Twice a year, in the spring and fall, more than 70,000 attendees come to High Point, North Carolina, to shop more than 12 million square feet in 186 buildings. (http://www.highpointchamber.org/hppc/index.html)

[14] The working team consisted of the Northern Economic and Social Development Office, the Northern Department of Industrial Promotion, the Northern Handicrafts Manufacturers and Exporters Association (NOHMEX) and the Faculty of Fine Arts, Chiang Mai University.

[15] In Thailand, the administrative structure can be divided into 3 levels: (1) the central or national administration; (2) the provincial administration; and (3) the local administration. Under the supervision and direction of the Cabinet, the central ministries and departments play major roles in policy formulation and implementation. Implementation and administration at the provincial level are in the hands of the regional offices of ministries and departments.

The provincial governor, who is appointed by the Ministry of Interior, is responsible for facilitation and coordination of the public programs of various government agencies. Lastly, the local administration, which consists of local officials who are directly and indirectly elected by the people, includes four types of governments: Provincial Administrative Organizations (PAO), municipalities, Tambon Administration Organizations (TAO), and special forms of city governance such as the Bangkok Metropolitan Administration and the Pattaya Administration. The local administration's major responsibility is to provide public services within its administrative area.

[16] The CEO governor as well as the COE budget was one of the results from the 2003 Government Reform scheme. However, these policies were cancelled after the military coup on 19 September 2006.

[17] The Provincial Comprehensive Plan employs a different set of classifications from the Urban Comprehensive Plan. They consist of 6 categories: (1) community; (2) rural and agriculture; (3) conservation of rural and agriculture; (4) open space for environmental quality and fisheries; (5) art and culture conservation; and (6) forest conservation.

[18] The field survey was undertaken between October - December 2006.

[19] The Wua Lai community is located right at the south of the old city. It was the traditional center of Chiang Mai silversmiths whose original roots were *Tai Kun* from Burma. Since early the 20th century, the community has produced and sold silverware ranging from household articles to ornaments such as bracelets, webbed belts, neck chains, food trays, boxes, and so on. During its prosperous era in the 1950s when Chiang Mai was first officially promoted as a tourist destination, there were around 200 silverware shops in the community. In the early 1980s, Ban Wua Lai declined when a new tourist center and center for the production of handicrafts was developed on San Kamphaeng highway by the Tourism Authority of Thailand and Chiang Mai Provincial Office. Traditional craftsmen of Ban Wua Lai were recruited to work in San Kamphaeng factories. In 2005, there were very few silver shops remaining in the Wua Lai community (Apawacharuth (2005)).

[20] San Kamphaeng has been promoted as a new, large-scale center for the production of handicrafts located outside Chiang Mai city since the 1980s.

[21] Interview with Thitirat Choeycharern, owner of Ngern Chiang Mai shop, Wua Lai road.

[22] Tha Phae pedestrian street was launched in 2002 by the Chiang Mai city government in collaboration with the business sector. It is a weekend market starting mid-afternoon on Ratchadamneon Road from Phra Pokkao Road to Tha Phae Gate. It is closed to traffic and serves as a place for stalls, all kinds of local events, and street performers such as musicians and puppeteers.

[23] Wat Ket community was a traditional community located on the east side of the Ping River. It is one of the largest communities in Chiang Mai, and was once very prosperous. The community was named after Wat Ket Buddhist temple, which is located at the center of community. It used to be an important port linking Ratanakosin, Chiang Mai, and neighboring countries. This port was no longer used after 1921 when the Chiang Mai train station was

built. Although the community has shrunk to a quiet residential neighborhood, its cultural prosperity remains visible through a number of beautiful traditional buildings. Many of them represent a great mixture of Lanna, Chinese, and Colonial styles. Religious buildings are scattered all over the community, including Wat Ket temple, Christian church and Muslim mosque.

[24] The old shophouse of Wat Ket received an award for best conservation in the vernacular community architecture category from the Association of Siamese Architects (ASA) in 2005.

[25] A residential area located on the northwest of Chiang Mai city. It was initially a low-density residential area. The area is mixed with small colonial style houses with front gardens and modern shophouses.

[26] The Nimmanhaemin Art and Design Promenade (NAP) was started in 2000 by the Nimmanhaemin Committee.

[27] San Kamphaeng was a center for the production of handicrafts located outside Chiang Mai city. The road was lined up with large-scale souvenir shops, factories, and some traditional villages producing handicrafts such as Ban Bo Sang (paper parasol) and Ban Ton Pao (mulberry paper). It was recognized as a counterpart of the Night Bazaar, the night market.

[28] San Kamphaeng was the hometown of the former prime minister Thaksin Shinawatra.

[29] Interview with Duangchan Apawacharuth, the director of Urban Development Institute.

[30] Interview with Nopawong Ratpatai, council member of Chiang Mai City Government, who owns a silverware shop in Wua Lai community and Sritoon Chaichaowong, Mayor of Ban Ton Pao City Government.

[31] Interview with Worrawimon Chairat, member of Wat Ket Conservation Club, as well as Thitirat Choeycharern, owner of Ngern Chiang Mai shop, Wua Lai road,

References

Apawacharuth, D. (2005) *Sustainable Cities in Chiang Mai: A Case of a City in a Valley*. Social Research Institute, Chiang Mai University

Chiang Mai City Government (2005) *Chiang Mai High Point Project. Civil Work Section*

Chiang Mai Municipality (2005) *Chiang Mai Municipality Development Strategy: Year 2005-2009*

Chairatana, P. (2006) *Conceptualizing Innovation Systems: Can Developing Economic Learn From Developed World?* (Ph.D. thesis of Department of Business Studies, Aalborg University)

Chulasai, et al (2000) *Economic Development in the Upper North Thailand*, Chulalongkorn Publishing

College of Arts, Media and Technology (2005) *The seminar report: Action Plan for City of Handicraft and Tourism Projects*

Department of Town and Planning (Chiang Mai Branch) (2006) *Chiang Mai Provincial Comprehensive Plan*

Kidokoro, T. (2005) *Sustainable Urban Regeneration and Local Governance in Regional Cities: Case of Kanazawa City*, in proceeding of International Workshop on Sustainable Urban Regeneration and Local Governance in Regional Cities, January 17-18, 2006, by Japan International Cooperation Agency (JICA) and Center for Sustainable Urban Regeneration, The University of Tokyo, Japan

National Economic and Social Development Board (2001) *The 9th National Economic and Social Development Plan (2002-2006)*

North Economic and Social Development Office (2003) *The seminar report: Development Strategy in Northern Thailand*

Roberts, P and Sykes, H. (2000) *Urban Regeneration: A Handbook.* Sage Publications

Pimonsathien, Y. (2006) *Cultural Resources Management in Historic Community in Bangkok: The Tha Tian Case* (A document for the Seminar on Urban Regeneration: Thailand and Japan Experiences, held by Faculty of Architecture, Chulalongkorn University, 24 October, 2006)

Pokapanichwong, A. (2002) *The Habitable Urban Revitalization: Nostalgia and the Middle Class's Construction of Social Space.* (Seminar Proceeding, Annual seminar of Princess Maha Chakri Sirindhorn Anthropology Centre. 27-29 March, 2002, Bangkok)

Satapitanont et al (2003) *Machizukuri.* Chulalongkorn Publishing

Chairat, W. (2006) *Baan Ta Wat Ket.* Chompu Printing and Copy

Sasaki, M. *Kanazawa: A Creative and Sustainable City* (available at http://www.ps.ritsumei.ac.jp/assoc/policy_science/102/10202.pdf (Accessed in 2007))

Shimizu, M. *Comparative Study on Regeneration of Cities in the Deindustrialized Age* (available at http://www.ps.ritsumei.ac.jp/assoc/policy_science/091/09104.pdf (Accessed in 2007))

Towiwat, P. (2006) *Local Economics Development through Handicraft Development.*

Acknowledgement

I would like to thank the Center for Sustainable Urban Regeneration at the University of Tokyo for providing a grant to conduct this research. My special thanks also extend to Dr. Panit Pujinda and Dr. Apiwat Ratanawaraha, Department of Urban and Regional Planning, Faculty of Architecture, Chulalongkorn University for their valuable comments on this paper, as well as Wittaya Duangthima, Department of Architecture and Environmental Design, Mae Joe University, for his great assistance during the data gathering process in Chiang Mai.

6. Spatial Regeneration and Beyond: Daegu, Korea

Shi-Chul Lee and Seungkeun Park

6.1 Introduction

Daegu, located in the southeast of the Korean Peninsula, boasts ample historic and cultural assets in and adjacent to the town. As agreed by many scholars, press, and ordinary people, not only has the city and region been distinguished in terms of Buddhism, but it has also been home to a number of persons of distinguished talents in many other fields: politics, economy and culture. However, recent decades have seen Daegu downgraded from Korea's top regional center to a mere declining local city. It has now even been argued that Daegu, which has already dropped to Korea's fourth largest city population-wise, might deteriorate even further.

Hot summers and the poor economy aside, recent tragic memories include a terrible gas explosion in 1995 and a deadly subway fire in 2003. Of course, Daegu is not alone when it comes to such familiar urban issues as downtown deterioration, sluggish local economy, job shortage, and environmental degradation. And while the nation has enjoyed various

Fig. 6-1. Gatbawee (4-meter Buddhist statue on Palgong Mountain) (Source: The City of Daegu, with permission)

fruits from Korea's world-renowned rapid economic growth in recent decades, it is also true that Daegu remains one of the least benefited, or even the most steadily-hit by the recent economic impasse, regional cities.

Yet, the city has had its defining moments and still boasts a great deal of remaining resources. Above all, Daegu has often been called 'the city of scholars.' An old saying goes that half of Korea's prominent scholars and high-level officials come from the Youngnam region; Daegu has been considered the center of Youngnam for at least the past 400 years. Daegu and its neighboring areas once took pride in having been the hometowns of three of Korea's presidents in the three decades since the 1960s: Park Chung Hee, Chon Doo Hwan and Roh Tae Woo. (It should be noted, on the other hand, that there has been criticism in regard to the issues of military dictatorship and human rights.)

Daegu has recently hosted some well-known international events such as the 2003 Summer Universiade and four important matches of the 2002 FIFA World Cup Games. It will also be the host city of the 2011 IAAF World Championship, one of the top international sporting events along with the Olympics and the World Cup. Will attracting global attention be enough to revive the old 'glory' of Daegu?

As a rather practical case report, this paper does not aim to present each of its points in depth as a purely academic article would intend to. We attempt to employ a conceptual frame, thereby answering the following study questions, at least in an exploratory manner:

1. In what fashion have Daegu's urban-scape and governance evolved?
2. How does the metropolitan region of Daegu fit in the overall frame?
3. How is physical spatial regeneration balanced with socio-economic restructuring in this region?

6.2 Background Information

Before moving onto the case of Daegu, it is worth briefly examining the local administrative structure in Korea and what has been going on in the nation, particularly with respect to the spatial de-concentration drive since the launch of the Roh Moo Hyun administration in 2003. Then, we will outline some basic features and the trends of Daegu's socio-economic structure.

6.2.1 Local Government Structure in Korea

As 'city,' a legal term in some cases as well as a traditional unit of planning analysis, does not wholly articulate what it could potentially embrace,

Table 6-1. Local governments in Korea (Source: Ministry of Government Administration & Home Affairs)

Regional governments (16)

- 7 Metropolitan Cities: Seoul (*TEUK-BYEOL-SI*), Busan (*GWANG-YEOK-SI*, hereinafter), Daegu, Incheon, Daejeon, Gwangju, Ulsan
- 9 Provinces (*DO*): Gyeonggi, Gangwon, Chungbuk, Chungnam, Jeonbuk, Jeonnam, Gyeongbuk, Gyeongnam, Jeju (Special Administrative Province)

Elementary level governments (230)

- 75 Cities (*SI*, autonomous under Province)
- 69 Districts (*GU*, autonomous under Metropolitan City)
- 86 Counties (*GUN*, autonomous under Province or Metropolitan City)

scholars have become familiar with 'city-region.' This concept refers to a network of cities or towns, forming a social, economic, and environmental system. Additionally, the concept of 'regional cities' refers to the central city of the city-region (Kidokoro (2006)).

Korea's government organization is a three-tiered system composed of national-, regional- and elementary-level governments, which combined produce 246 *local* governments as of June 2007, as shown in Table 6-1.

The City of Daegu is one of Korea's seven *regional* governments, which might cause confusion to international readers. In this paper, we interchangeably use the terms 'Daegu' and 'the City of Daegu,' which both indicate the limited geographical boundary or local administrative jurisdiction. However, the 'Metropolitan Daegu region' refers to the City of Daegu and its neighboring areas combined, including several other cities and counties.

6.2.2 Korea's Recent Decentralization Challenges

The growing concerns of over-concentration in Seoul are not new in Korea. Population distribution, an undoubtedly important indicator for spatial concentration, reveals the reality of the situation, as shown in Fig. 6-2. The Capital Region houses more than 21 million inhabitants, a little less than a half of the total population, which is already a large number considering that only 20% of Koreans lived in the Capital Region a half-century ago. Future predictions show that the non-Capital regions will continue suffering in the coming decades as well. The number is significantly large when compared to figures from other OECD countries such as Japan, France and England, whose population share of the capital region are 31.9%, 18.5%, and 11.8% respectively.

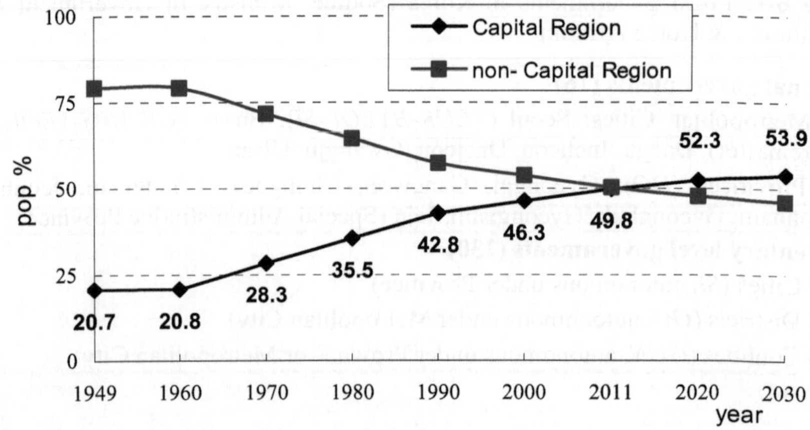

Fig. 6-2. Predicting population change in the Capital/ non-Capital region (Modified from National Statistics Office (2005))

In addition to population dispersal, Korea's spatial disparity in terms of economic values and resources is even more serious, illustrated clearly by a glimpse of existing conditions (KRIHS (2000)); (Ministry of Construction and Transportation (2002))

1. The area of Seoul is 606 km^2, only 0.6% of the nation's total land. Seoul's aggregated land price is believed to exceed 37% of the nation's total. The Capital Region, covering 11,745km^2, or less than 12% of the nation, claims about 66% of total bank deposits and 62% of total loans.
2. The Capital Region is home to approximately 55% of the manufacturing firms, 66% of the regular four-year universities, 73% of all R&D institutions, 77% of Korea's venture businesses, and 84% of the national government agencies and organizations.
3. About 68% of the total number of, and 53% of the total amount of, the Foreign Direct Investment (17,802 cases, US$ 31 billion) during 1962-2003 have been funneled into the Capital Region.
   It is not surprising that the headquarters of 95 of the country's 100 largest private firm headquarters and 80% of the state-controlled public corporations are located in Seoul and its vicinity.

Since its launch in 2003, the Roh Moo Hyun's Participatory Government has employed a series of unprecedented policy measures for decentralization and balanced national development as an on-going reform strategy. The most distinguished campaign was, arguably, the debate revolving around the construction of the Administrative Capital City in Chungnam

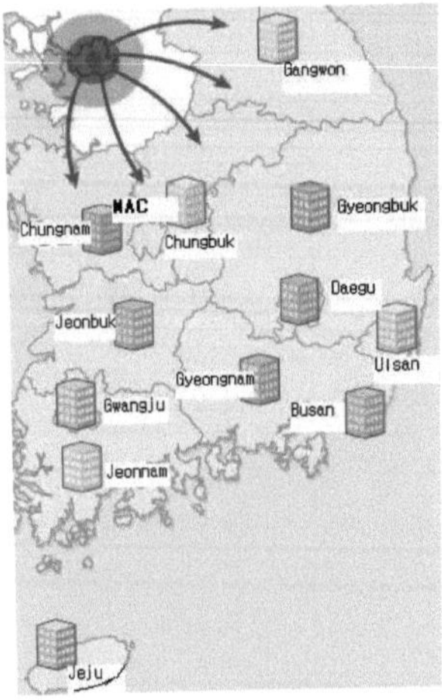

MAC: Multi-functional Administrative City (Sejong)

- Location: Yeongi, Chungnam Province
- Population goal: 500 thousand residents
- Year to be completed by: 2012
- Offices to move: 12 Ministries, 4 Agencies

Inno-Cities: relocation of major public bodies

- Receiving Regions: 12 Metro Cities & Provinces
- Employees to move: 32,000
- Year to be completed by: 2012
- Moving entities: 175 out of 345 state-owned corporations (not Ministries)

(Daegu is scheduled to house the Korea Gas Corporation & 11 other public bodies, and Gyeongbuk will be home to the Korea Highway Corporation & 12 major public agencies.)

Fig. 6-3. Korea's de-concentration plan (Modified from Ministry of Construction & Transportation)

(South Chungcheong) Province. In the process, however, the relevant law was declared 'unconstitutional' in 2004; the whole project has been downgraded to the construction of a 'Multi-functional Administration City (MAC),' now called *Sejong City* (Lee (2007)); (Multi-functional Administrative City Construction Agency).

Another noteworthy project concerns the construction of 'Inno-cities' scattered around the nation, where major state-owned corporations are to be relocated in coming years, as shown in Fig. 6-3. The region-by-region relocation of the public organizations has been decided. The plan, which received clearance from the National Assembly in 2005, after surviving accusations that it is unreasonable and an astronomical waste of money, is a substantial and unprecedented attempt to balance nation-wide development. Apart from the already intense cost-benefit efficiency dispute, this initiative is comparatively on track, in terms of goal attainment and localities' willingness. The move could start as early as 2007 and is scheduled for completion by 2012, although both of the grand projects still face considerable barriers such as enormous costs and the probable negative impact on, and thereby political resistance from, the Capital Region (Lee (2007)).

6.2.3 Daegu, Now and Then

Daegu is located inland, southeast on the Korean Peninsula. The city is 295 kilometers (184 miles) and 122 km (76 miles) away from Seoul and Busan, respectively. It takes only 99 minutes from Seoul by KTX, Korea's

- Population: 2.51 million persons (2007. 6)
- Area: 885.6 km^2
- Population density: 2,873 persons/ km^2
- Districts (autonomous local governments): 8
- City government budget: US$ 4,061 million (2007)

(City & 8 Districts combined: US$ 5,558 million)

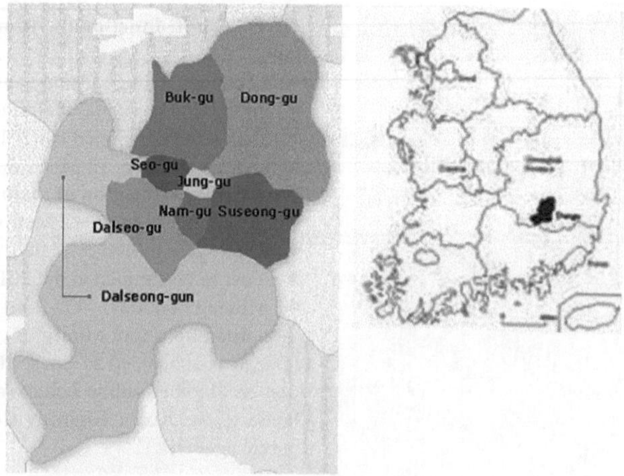

Fig. 6-4. Daegu in map and numbers (Source: The City of Daegu)

high-speed train system, which was launched in 2004. The City of Daegu covers an area of about 885 km^2, or 0.9% of the nation's land. The number of the residents is approximately 2.5 million, which is 5.3% of the nation's total population of about 48 million.

The city's ancient, but the still-used, name Dalgubeol appeared in history more than a thousand years ago during the Shilla Dynasty era. 'Daegu,' literally meaning a 'Great Hill,' has not only changed its name and physical shape several times, but has also transformed itself from a small castle into an agricultural capital of the region, and finally into a modern city.

Since the establishment of the administrative headquarters of the Youngnam region (the current Busan and Gyeongnam areas included) in 1601 during the Chosun-Dynasty, Daegu has remained the core of the whole region for about 400 years. Arguably, the first modern city of this area was born in 1895 when it was given the official title of 'Daegu *Bu*,' and it remained the central city of the whole region. After Korea's liberation from Japanese rule in 1945, Daegu became a *Si* (City) and the capital city of Gyeongbuk Province. During the Korean War in the 1950s, the city was one of the few surviving major cities, along with Busan.

In 1981, the city was promoted to a Metropolitan City, administratively separate from Gyeongbuk Province, incorporating many bordering areas into its territory. In tandem with the revival of Korea's local autonomy over a period of 30 years, the first election for the city council took place in 1991, and the first mayoral election occurred in 1995, a year which also witnessed a significant annexation of the entire Dalsung-gun area, consisting of 431 km^2.

Population growth in Daegu has significantly slowed down for the past decade, compared with the 1960s, 1970s, and 1980s. (The considerable increase of population in 1981 and 1995 are apparently due to the administrative annexation as earlier mentioned.) In fact, since 2005, the net population of the city has begun to decrease for the first time ever, and is projected to continue for a while.

It is largely believed that the development of Daegu and many other regional cities depend on outside investment, particularly from foreign direct investment (FDI). However, the truth is that not many regional cities have benefited from FDI. During 1962-2003, about 68% of the total number of, and 53.1% of the total amount of, FDI has been concentrated into the Capital Region (KRIHS (2000)); (MOCT (2002)). The local share has been negligible, except for a few large national project areas such as Ulsan and Pohang. Out of the 16 metropolitan-level governments, Daegu has received the lowest amount of FDI for the past three decades.

A quick review of the Gross Regional Domestic Production (GRDP) statistics shows a more vivid reality for Daegu. In 2006, the GRDP of Daegu, with 5.3% of the nation's population, totaled 26,046 billion won (US$24.8 billion), only 3.3% of the nation's total 815,289 billion won (US$774 billion). The combined GRDP of the Capital Region (Seoul, Gyeonggi, and Incheon) keeps increasing compared with the rest of the nation, comprising about 47.4% of the nation's total GDP. When it comes to the GRDP per capita, Daegu (about US$10,600 per person) ranks the 16th among 16 Metropolitan Cities and Provinces (National Statistics Office (2006)).

6.3 Urban-scape and Planning of Daegu

Beginning with the brief overview of how Daegu's urban spatial patterns have evolved, we examine how and by whom the city's urban expansion development has taken place. Then, the overall system and responses of national and local planning will be outlined.

6.3.1 General Pattern and Inner City

Daegu is considered a comparatively old city, and its modern urban pattern originates from less than a century ago. It is widely believed that modern city planning in this city was introduced as early as 1932 under Japanese occupation. Since then, Daegu has been largely a traditional mono-centric city. Unlike Korea's other major cities such as Seoul and Daejeon, Daegu survived the Korean War in the 1950s without being bombed or occupied. Old street patterns in the city center have remained mostly intact; back streets around the city are also scattered in and adjacent to the downtown area.

Since the mid-1960s, two distinct patterns have been observed: inner-city redevelopment and outward expansion of suburban areas. Fig. 6-5 shows the sprawling pattern of the city's expansion during 1987-1995, which showed one of the greatest area- and population-wise growths in this city.

Due to the great development pressure in the 1980s and 1990s, a series of growths in built-up area have occurred in tandem with the expansion of the administrative boundary as previously mentioned, even beyond the outer edge of the Green Belts, a three-decade-old staunch policy measure to control urban growth in Korea.

As shown in Fig. 6-4, about a 40% increase of built-up area was seen during the 1987-95 period alone. This sprawl and strip development pattern occurred mostly along the major arterials, and toward the Palgong mountain areas, costing a huge loss of open spaces, and an encroachment of agricultural land and other green space. A recent city master plan suggests that the city will evolve with into two urban centers and five sub-centers. The urban centers are the old town in the Jung-gu area and East Daegu new town, while the sub-centers include Chilgok, Anshim, Seongseo, Dalseo, and Hyunpung.

With some deviations, major urban redevelopment projects have been implemented mostly by the private sector, as in the case of transforming urban brownfields or old school sites into high-rise apartment complexes.

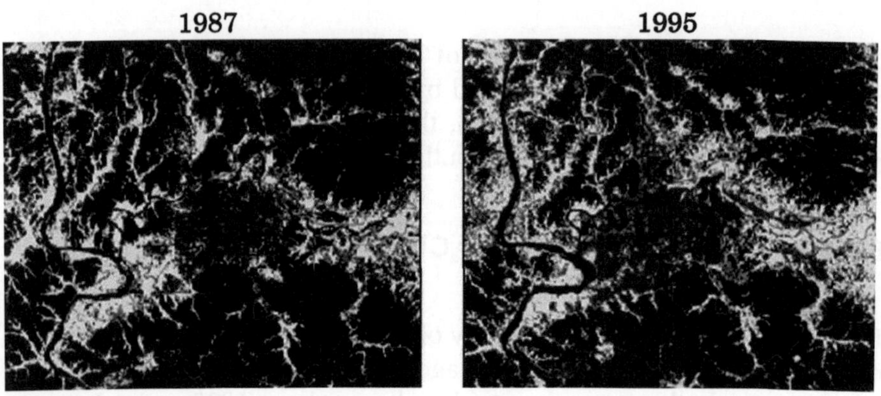

Fig. 6-5. Sprawl pattern expansion of Daegu (Source: KRIHS (2005))

Chimsan-dong, a typical redevelopment site, which used to be a manufacturing venue for Cheil Industries, a parent company of Samsung, is an example. Located north of the Daegu train station, the whole area was Daegu's first industrial complex, housing major textile firms. Due to the continuing development pressure, most companies have moved out to the suburbs, or even further into neighboring areas such as Cheongdo and Gumi. The plants have been replaced by business and residential facilities, mostly by big-box discount shopping centers and massive high-rise apartment clusters. Although the public sector's role is believed to have been largely limited in the development process, the local government came into play by taking initiative in some redevelopment projects, especially since the 1990s, as shown in the Taepyung-ro redevelopment project in the downtown area.

6.3.2 Suburban Development

In the 1980s and 1990s, there was significant annexation of suburbs into the City of Daegu, especially coupled with political and administrative rearrangements. Fig. 6-6 shows some of the significant physical expansion of the city's administrative boundaries over the past few decades.

1963-81 | 1981-95 | 1995-present

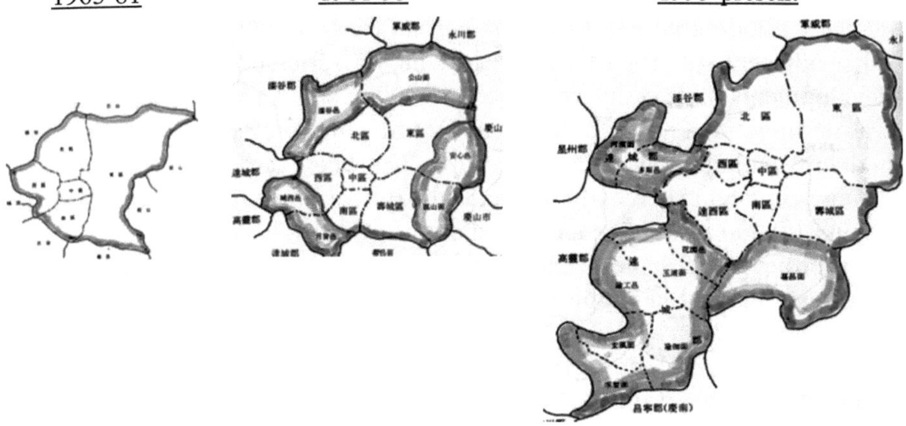

179 km^2 | 455 km^2 (Anshim, Chilgok, Gongsan, Seongseo & Wolbae annexed) | 886 km^2 (The entire Dalseong–gun annexed)

Fig. 6-6. Expansion of Daegu's administrative boundaries since 1980s (Source: Daegu Metropolitan City (2005), The City of Daegu)

- Location: 15 km north of city center
- Annexation year: 1981
- Current population: 100 thousand residents

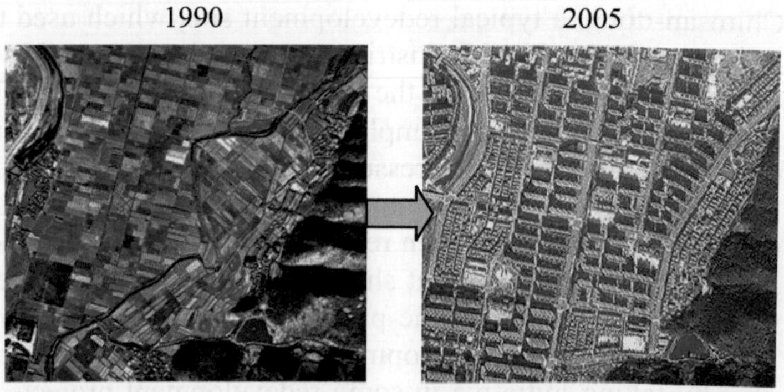

Fig. 6-7. Suburban development in Chilgok: Rice fields to apartment complexes (Source: Drawn from The City of Daegu 2005)

These additional lands have been readily used for further development, in many cases, in a sprawling fashion. During the 1990s, unmanaged macro-level planning or legal intervention by the national government aside, a number of sprawling cases have been witnessed in the Daegu area, causing longer commuting distances, traffic jams, and environmental degradation in traditionally green areas, such as the Palgong mountains. These large-scale development projects in the suburbs have been initiated by the public sector in the early stages but were later implemented mainly by market forces.

The Chilgok area is a typical suburban transformation case, along with other adjacent suburbs of Daegu, in which a series of massive residential developments have occurred since the 1990s. About 15 years ago, this area was nothing but a northern fringe of Daegu, consisting mostly of rice fields and low mountainous areas. The area has been through some dramatic changes in land use, mostly due to high-rise apartment complexes, housing about 100,000 new residents, as shown in Fig. 6-7. A series of development projects that have occurred during the past couple of decades in the area have resulted in a typical bed town, and bed town alone, of the mother city Daegu.

Chilgok is not alone in this nature of suburban development in the region; other major projects took place in such areas as Wolbae, Seongseo, and Dalseong area, all of which were annexed into the City of Daegu in the 1980s and 1990s.

6.3.3 Planning Mechanism and Responses

Korea's spatial planning had been governed mostly by the National Land Use and Management Law (NLML) and the City Planning Law (CPL) until 2002. The latter was applied exclusively to those areas designated as 'city planning areas,' which cover most major cities and built-up areas in small and medium-sized cities. The NLML, initially enacted in 1972, was the general framework for the nation's land use, focusing more on the rural areas left out of the city planning areas. It is largely believed that local government's role in planning was fairly limited in this structure (Lee (2000)).

In 2003, the two major legal bases were merged into the National Land Use and Planning Law (NLPL). The new law suggested a significant change in the planning principle: 'planning first before development.' It also required region-wide planning for major metropolitan regions such as Daegu. A growth management requirement, 'concurrency,' was also officially introduced in the new 2003 law. Many of these changes have affected local planning, particularly for metropolitan areas.

Based on the Constitution and the Comprehensive National Territorial Development Planning Law, the central government is required to build comprehensive and fundamental long-range plans governing all lower-level plans. Concerning not only land use and development but also population, infrastructure, industries, and environment, a 10-year national plan has been set up and implemented three times: the first one for 1972-81, the second for 1982-91; and the third for 1992-2001. The Fourth Comprehensive National Territorial Plan (2000-20) is now in use (www.moct.go.kr).

Pioneer modern city planning in Daegu was introduced as early as 1922, but the Daegu City Master Plan, a 20-year long-term plan as in the current form, began in 1984. The City Master Plan and the City Management Plan are the two major planning documents in this city. The former, as a 20-year comprehensive plan, requires a mandatory update every five years. It is a 'big picture' plan covering a variety of planning fields, not only spatial but also socio-economic. Daegu City, more specifically, the Urban Affairs and Construction Bureau, just finished establishing the 2030 City Master Plan last year. Based on the master plan, a more detailed City Management Plan is also to be established every five years (Daegu Metropolitan City 2005).

Gyeongbuk Province, the parent province of Daegu, has its own Provincial Comprehensive Plan in accordance with the relevant law and the national comprehensive plan. As for the Metropolitan Daegu region, however, it was not until February 2003 that many rural areas adjacent to Daegu were required to adopt a city plan. Naturally, there has not been any systematic or meaningful relationship between the Daegu's master plan and provincial plans up until recent times.

An evolution of region-wide spatial planning was launched in 2005, which was a requirement based on the aforementioned new legislation,

NLPL. The Metropolitan Daegu Regional Plan covers the City of Daegu and seven localities of Gyeongbuk, all adjacent to the city. Initially intended to cope with the 30-year old Green Belt issue, the region-wide plan embraces general land use planning, population distribution, natural resource protection, various region-wide infrastructure, etc. Largely considered the very first outcome of the joint spatial planning efforts between Daegu and Gyeongbuk, there are great expectations, but how it will turn out remains to be seen.

In addition, it is expected that there may be other considerable changes in the local planning system. Local voices have stated that major planning approval authority and discretion, which is currently held by the central government, can and should be devolved to metropolitan cities and provinces. Then, local governments would be able to finalize any physical urban restructuring with their councils' agreement, and without the approval from the Central City Planning Committee in Seoul. Additionally, in compliance with the 'concurrency' requirement from the 2003 new national planning law, local governments have been also required to harmonize planning and major projects. Some specific policy requirements have been already been articulated[1].

Daegu's transportation network, both city and region-wide, deserves attention, too. On top of the fact that the city as a whole has benefited from its location, Daegu also boasts an excellent regional road system, together with seven major expressways and easy access to major ports such as Busan and Pohang, both only an hour away. The city and adjacent areas' everyday transportation is covered partly by semi-public buses and subway systems. Two subway lines are already in operation, while Line No. 3 is under construction and No. 4 planned. The only weakness in light of transportation lies with the comparatively small Daegu International Airport. Currently, a plan for a new regional international airport, as a bold policy agenda, has been raised and is continuously being articulated in partnership with other neighboring localities, even the Busan region.

6.4 Daegu's Urban Regeneration and Beyond

Daegu is certainly not alone when it comes to downtown deterioration; this phenomenon has appeared in many other Korean cities, let alone major Western cities, over the past decades. The issue of Daegu's urban regeneration is discussed based on the current pattern of the city center's decline and the overall disproportional development arrangement. Physical development, which is to be balanced with socio-economic reshuffling, leads to the significant issue of regional economic integration.

6.4.1. Existing Conditions: Decline and Disparity

While Daegu has a greater amount of land available for development than Seoul and Busan, there exists substantial disparity in the development density among the districts in this city. The city has become polarized into overdeveloped districts and underdeveloped areas.

As Daegu's CBD moves and extends toward the east and the south, forming new centers in the East Daegu station area and the Suseong area, Daegu's old city center (Jung-gu) has been continuously declining, in terms of both residents and businesses. Firms have vacated downtown buildings; local small businesses have moved out. Where the original parent market of Samsung (Samsung Sanghoe) was established in 1938, one can find only a number of small shabby tool stores along the street. The nightscape in downtown Daegu has usually been characterized as being dark and empty.

Overdevelopment in some areas of Daegu is another problem. The commercial districts around east Daegu, the new town, are heating up by private enterprises, maybe in an extreme fashion. Such a situation has resulted from allowing mixed-use high-rise development (e.g. skyscraper apartment complexes along the major arterials and in commercialdistricts). In Suseong-gu, for instance, more than 100 development projects were approved by the end of 2005, and this trend continued in 2006 (Daegu-Gyeongbuk Development Institute (2006)).

The East Daegu station area, the so-called entrance to Daegu, has not been given sufficient priority for further development. Some maintain that the KTX station area should have been greatly reorganized in 2004 when KTX started its service, a reference to the lack of commercial development, urban infrastructure, and efficient transportation network within the station area. Just as in the old central area, the city's overall investment in the area has fallen considerably short and has been delayed.)

Fig. 6-8. The East Daegu Station Area (Source: The City of Daegu)

Another indispensable citywide issue concerns the existing industrial complexes that have deteriorated. For several decades since the industrial complexes were established in Daegu, they have been home to approximately 18,000 employees as well as 2,000 small and medium-sized firms. Yet, due to increased land prices, despite the age of the facilities, it has become difficult to keep or enhance the overall productivity in these areas. In addition, situated so close to the city center, the complexes not only function inefficiently but also keep spoiling the beauty of the cityscape and damaging the environment.

6.4.2 Efforts and Prospects for Urban Regeneration

It makes sense to argue that it is still early for Daegu to become 'regenerated,' in part because the city and the adjacent areas have never fully 'flourished.' Unlike many Western cities, the physical growth of the city has not matured enough, so the transformation of the city is still an ongoing process. Another argument goes that considerable 'growth' or 'development' should come first before 'growth management' or 'urban regeneration.' While admitting the reasoning, we examine some noticeable efforts and a prospect for regeneration, with emphasis on the physical aspects of some specific areas.

The situation in the old city area has not greatly improved, and private enterprises are not happy to participate in redeveloping the area. Yet, there is a hope for the old town. Most recently, some downtown revitalization projects have been initiated by City Hall and are supported by the business community, many of them not entirely with local affiliation. This is encouraging, in part because statistics reveals that the daytime population in the city center still remains high.

Redevelopment in the old central area is expected to begin as soon as possible in order to revitalize the whole city. In fact, the City of Daegu has already assigned 66 districts to a maintenance project zone in Jung-gu alone. However, it remains questionable whether the schemes will be

Table 6-2. Residents and daytime population in Daegu's city center (Source: National Statistics Office (2005))

	Number of residents	Number of daytime population	Percentage (d/r*100)
2000	84 thousands	142 thousands	168.1%
2005	73 thousands	136 thousands	186.1%

properly carried out considering that the city government needs to rely heavily on private enterprises for the entire project. Notwithstanding the goal of expanding the role of the public sector, not many concrete plans have been specified yet. Inducing investment from businesses also faces a great deal of difficulties. It is long overdue that City Hall take the lead by providing creative and diverse incentives to attract the market's cooperation. City Hall also ought to positively scrutinize the overall entry barriers and other relevant regulations on redevelopment.

It is also to be noted that the Daegu region is well known for its historical and cultural assets. Many cities around the world have carefully preserved their historic heritage. In modernizing a city's central area, those historic monuments must be well conserved. Small iron sculptures in old streets of Stockholm attract foreign tourists, and many people go to Kanazawa in Japan to visit the glasswork museum – history can bring vitality to a city. Daegu's historic sites and cultural resources ought to be preserved for economic rationales, as well as social reasons. In addition, art and culture can significantly contribute to revitalizing the city. Monuments, museums, libraries, art performances, and various exhibitions would fill the city with charm. Aiming to become a city of art and culture, especially for musicals, the city has been planning to considerably increase the art and cultural infrastructure, which should continue in the future.

The physical regeneration of a city can never be enough without design consideration. Also in Daegu, an urban design committee has been organized to create an image of Daegu, looking forward to next 100 years. The committee sets and oversees the basic directions for plans such as city revitalization, river conservation, and industrial complex maintenance.

In addition, the Daegu city government has promoted a number of significant projects for regeneration, such as the citywide Green Daegu program and Fence Elimination campaign; the city government created a budget of 10.6 billion won for these two programs. It has been maintained that the overall temperature, and even the city image itself, has been vastly improved or changed, thanks to a wave of large-scale 10-million tree planting drives (arguably, Daegu is not Korea's hottest city any more.) The Fence Elimination project is a national benchmarking case, for which more than 17 km walls or fences from about 400 houses and buildings in the city have been torn down over the past 10 years. A five-year plan on air improvement is also in progress. The city is making considerable efforts in expanding the infrastructure for water quality improvement, which in turn appears to play a decisive role in promoting Daegu as a sustainable city and upgrading its living conditions.

6.4.3 Efforts for Economic Restructuring in the Region

The economic features presented in this paper with respect to the Daegu region are two-fold: dealing with existing manufacturing bases within the city and creating entirely new grandiose projects.

Apparently, the manufacturing industry, 21.7% of the city's GRDP, keeps declining. Prevailing beliefs hold that the local government's leadership in restructuring Daegu's industry has not been effective for decades. The city's major industry is still textile, and local efforts have recently focused mainly on retaining the existing labor-intensive manufacturing firms. On the other hand, knowledge-based industry in Daegu accounts for much less than half that of Seoul, and even smaller than the national average. Daegu has been the lowest in the nation when it comes to FDI inducement.

On top of that, deteriorating industrial complexes and declined productivity, due to the changed industrial structure, have caused trouble in terms of land use, as well as business activities. Ideally, the existing enterprises in the complexes should eventually be moved out to newly established industrial sites at the edge of, or right outside of, Daegu. In fact, the Daegu city government is planning to have the current sites 'reborn' and invite high-value-added firms and businesses.

All in all, Daegu's current efforts to restructure the overall industry appear to be centered on a few grand projects for the provision of research and industrial land, which are not new. In retrospect, big projects such as construction of the Wicheon Industrial Park and the Bongmu Apparel Complex have been unsuccessfully attempted since the 1980s. The effectiveness of the 'Milan Project' in the 1990s to support local textile industries is still in dispute.

To address the industry restructuring issue, several on-going grand projects are marked, and two, among others, are particularly noticeable: the Daegu Technopolis and the Culture Industry Cluster.

The Daegu Technopolis, located in Hyunpung, southwest of Daegu, and covering an area of 1,800 acres, is meant to be a future center for research, education, and housing. With its master plan authorized in 2006 (the construction already began in 2006 in partnership with the Korea Land Corporation) the project is scheduled to be completed by 2015, at a cost of approximately US$2 billion, or 1,900 billion won. It would be home to some major institutes such as the Daegu-Gyeongbuk Institute of Science & Technology, an ETRI branch, the National Science Center, and hopefully the Kyungpook National University College of Engineering and the Medical School.

The second major project, the establishment of the Culture Industry Cluster, has been planned in tandem with IT resources and is based on Daegu's abundant cultural assets. Based on the Daegu Culture Industry Development Master Plan authorized in 2004 by the Ministry of Culture and Tourism, a redeveloped college site in the city center is being utilized to construct theme parks, traditional folk villages, and various exhibition and convention facilities. It is also expected to be completed by 2015; the expected costs could reach US$1.4 billion (1,300 billion won).

Last but not least are the economic issue concerns the KTX, which started service in 2004. There have been two types of impacts on this region. The positive effects are more shoppers from neighboring areas, quicker and more convenient trips to Seoul, greater opportunities for leisure trips, and the expected development of railway station. On the other hand, the city is not happy because there are negative effects such as so-called the centrality shift to Seoul and even to Busan and a significant increase in medical (38% of KTX trips) and other high-value-added trips to the Capital Region (Lee & Kwak (2005)).

No doubt the East Daegu station, as the KTX center of the city, has become the main doorway to Daegu. The areas around the East Daegu station, however, still have not changed much from their unclean and aged past, which keep generating negative impacts on the city. Therefore, the whole station area needs to swiftly transform into a new and fresh face of Daegu filled with vitality. While regeneration of the old center should be based on culture, tradition, and administration, the emphasis in the East Daegu station area should focus on business and commercial uses so that it can be functionally harmonized with the other side of the city.

The total estimated cost for these major projects amounts to 13 trillion won, or US$17.3 billion. The centerpiece of the potential appears to be the region's abundant human resources, e.g., about 50,000 annual graduates from 33 colleges and universities in the region. In fact, Daegu City has recently been developing more specific knowledge-based, and also business- and industry-oriented, projects. If these projects are well integrated and successfully achieved, Daegu could eventually be called an 'Asia's Brain Capital' as City Hall intends (Daegu Metropolitan City (2007)).

6.4.4 The Issue of Economic Integration of Daegu-Gyeongbuk

As is geographically obvious, Daegu had been a part of Gyeongbuk Province for the past 100 years. However, since Daegu was promoted to a metropolitan city in 1981, the city has been separated, not only administratively but also economically in a sense, from Gyeongbuk Province.

The two local governments began to independently form plans and promote development in most fields such as regional and urban planning, industry establishment, and tourism development. Moreover, the two jurisdictions have focused more on competition than on cooperation with each other by adopting different policies. As a result, they have achieved their own developments, but they also have occasionally built an antagonistic relationship with each other, without fully making use of their potential. Many local residents believe that the separate administration systems have frequently created adverse effects such as unproductive competition and overlapping investment, and resources were wasted due to ineffective operations.

Administrative dependence aside, the two regions are closely interconnected in light of their economies and everyday life. As mentioned earlier, Daegu's relative level of GRDP per capita is 62.6 (Korea=100), the lowest in the nation; on the other hand, consumptions amounts to 98.3, ranking the city fourth. The yearly household consumption expenditure per capita of the city amounts to US$8,280, while that of Seoul is US$9,986 and the nation's average is US$8,417. It is natural to conclude that Daegu, as is the case with Korea's other two major struggling regional cities, Busan and Gwangju, depends considerably on neighboring localities, and non-Daegu residents' income.

Yet the two regions share economic difficulties, too. For example, the relative decrease in Daegu's overall income and economic value has kept transferring difficulties into neighboring Gyeongbuk province. On the other hand, the province's suffering, not least by recent deregulation in the nation's Capital Region, also causes severe crisis in not only Gyeongbuk but in Daegu. Other factors have also contributed to the current crisis of Daegu and Gyeongbuk. Traditionally believed to be very conservative, the two regions have shown great reluctance in transforming their industrial structure, whereas some other regions in the country have dared to change or reshuffle their main industries (Kim (2005)).

Although the two local governments established different administrative practices since their separation, it would be fair to say that they have shared the same character, dialect, and culture among local residents. Recent statistics show that the number of commuters leaving Daegu in the morning overwhelms that of those entering from the neighboring areas (about three to one). Approximately 68,000 people commute to firms and colleges in Gyeongsan; there are about 20,000 commuters for Gumi (mostly firms, some colleges) and Chilgok (firms, businesses). This of course indicates that Daegu's jobs and economy in general depend largely on adjacent areas in Gyeongbuk.

Recently, there has been a wide sense of accord that the two neighbors seek 'real' cooperation. The movement for the economic integration of Daegu and Gyeongbuk is of great significance in order to overcome the troubles and gain a foothold in a self-supporting economy. The administrative integration, however, eventually requires legal support from the central government and takes a long time; hence, it has been widely agreed that economic integration should come first. In fact, Daegu and Gyeongbuk Province exchanged an MOU in 2006, reaching a basic agreement on economic integration, which will hopefully be a significant steppingstone for economic recovery in the whole region. Since then, a joint Daegu-Gyeongbuk Economic Integration Forum has been founded for necessary conferences and seminars on the integration project. In addition, the Economic Integration Promotion Committee has been created, in which local professionals and public officials are to develop various strategies and action plans. Based on recent discussion, some promising projects have been identified and are under serious consideration or in practice:

- Establishment of a mobile test bed: settled to be built in Gumi, Gyeongbuk
- Expansion of the EXCO convention center: Gyeongbuk's investment of 10 billion won planned
- Inducement of foreign investment and other international cooperation: jointly activated, planned, and implemented, for example, the NEAR operation (Gyeongbuk-initiated international organization)
- City-country cooperation: Gyeongbuk's agricultural products promoted to be sold in Daegu
- Transportation and tourism: region-wide smart card compatibility, tourism promotion (KTX)
- Others: university integration in the two regions, creation of a united database for the economy

It is questionable to what extent they can share the same economic policies and goals as initially expected, given that the two governments operate under separate administrations; however, if they succeed, Daegu and Gyeongbuk will have achieved much more than just economic integration for the first time in Korea.

6.5 Governance of Daegu Region

While discussing the spatial regeneration and economic progress of Daegu and its neighbors, it is also worth exploring the issue of local governance, which just about everybody has talked about. How and by whom are major policy and planning decisions made in this region? Is (local) government still prevailing or has civil society actually come into play?

6.5.1 Evolving Issue of Governance

The centuries-long centralization in administrative authority in Korea has only begun to change over the past decade. Although some achievements of the current Roh administration for administrative devolution to local governments deserve appreciation, it still remains to be seen how far these devolution efforts can possibly reach (Lee (2007)). With that said, we turn to the issue of local governance. Particularly in Korea since 1995, local governance has witnessed a great deal of change, as shown in Table 6-3.

Undoubtedly, urban regeneration has to be planned and implemented together with the decentralization of administrative power from the central to local governments so that it can be successful in resolving urban issues. And citizen participation is indispensable to urban regeneration (Onishi, 2007). Contemporary society, based on ongoing development of information technology including the Internet, has encouraged an upgrade and diversification of all fields of society. Further, the role of NPOs and NGOs has seriously expanded and intensified, although many self-governing bodies in Korea are not yet sufficiently functioning to aggregate and articulate public ideas and opinions, and the local council members have not completely and appropriately represented the general public.

The civil society in Daegu, and perhaps in many other Korean cities, appears to have played a very limited role in the overall process of spatial regeneration. It has long been suggested that the basic concept and role

Table 6-3. A paradigm shift in Korea

	Before 1995 Mayoral election	After Mayoral election in 1995
Focus	Service provision	Economic development
Planning emphasis	Supply-side, 'Hardware' provision	Demand-side, 'Software' planning
Methods	Hierarchy & control	Partnership & cooperation
Mode	Managerialism	Entrepreneurialism
Key players	Local government	Local government & civil society

of local government and its function thereof should be changed in Daegu, too, in order to become a regional hub that progresses toward the future (Kim (2003)). The reliability and effectiveness of policies can be enhanced by fully allowing the interested parties and residents to take part in the decision-making process.

As for both economic development and administrative reform, privatization and commission of public services, together with inducements of private investment, should be constantly achieved by means of reinforced partnership between the government and the market sector as well as NPOs and NGOs. In order to improve the existing centralized structure of the administration, systems such as the UK's CCT (compulsory competitive tendering) could be further sought out and put into practice.

6.5.2 Governance Participants in the Region

It is not surprising to argue that government cannot, and should not, be the only authoritative institution and that even 'governing without government' could exist in certain circumstances (Peters (1996)); (Rhodes (1996)); (Stoker (2003)). In a sense, local communities can be run by integrated mechanisms through a mixture of entities such as the general public, local governments, NGOs, universities, and private firms.

It goes without saying that the priority in the decision-making process should be given to the public, either as individual citizens or groups. Recent institutional changes can be directly utilized by residents in a positive fashion. For instance, the Citizen Referendum Act, enacted in 2004, has already been brought into play in three actual cases in the nation. The Citizen Recall Act passed in 2006 also provides one useful checking instrument for local communities; the act is already being experimented with in Hanam, Gyeonggi --- against the city's mayor. While these legal devices could invoke confusion and conflict in the near future, they are expected to help build better local governance and monitor the overall process of decentralization as well, thereby ushering local governments into uncharted territory (Lee (2007)).

Aside from local governments, there are various community organizations in the Daegu region, many of which are not entirely unique in this area. The leading NGOs that have come into play would be, among others, the Daegu Citizens' Coalition for Economic Justice and the Daegu People's Solidarity for Participatory Democracy. These organizations armed, arguably, with democratic morality and sound professionalism, but not with great financial resources, have greatly contributed to the general process of important decision-making in this region. Not surprisingly, fierce conflicts and candid disagreement have occurred in significant policy processes, for instance in the Apsan Tunnel Project in 2005, a major construction development project involving possible traffic solutions and environmental costs.

Some organizations are exclusive to this region. Yet these NGOs are mostly affiliated administratively or financially with the city or provincial government. Among others, we identify Local Agenda 21 (the Green and Clean Daegu 21 Promotion Committee), the Daegu-Gyeongbuk Regional Innovation Council and the I Love Daegu Citizen Conference.

The Green and Clean Daegu 21 Promotion Committee was organized based on the "Daegu City Environment Ordinance" (October, 2006). The committee set up four sections (Making Clean Daegu, Making Green Daegu, Making Healthy Daegu, and Together with Daegu), which are followed by 25 action principles and 275 practice principles. The activities include the following:

- Establishment and assessment of the plans for Local Agenda 21
- Education and promotion of Local Agenda 21
- Integration and cooperation with institutions involved with Local Agenda 21
- Research on policies for environmentally sustainable development

The Daegu-Gyeongbuk Regional Innovation Council, established after the launch of Roh administration, is a statutory institution by "The Special Law on Balanced National Development," based on which similar organizations were set up nationwide. It plans and regulates the major projects related to the regional innovation plan and balanced development. The Council aims to accumulate social capital, effectively form local governance, and establish the best regional innovation system (RIS) in the nation by 2012. The mayor of Daegu and the governor of Gyeongbuk Province, based on recommendations from universities, research institutions, corporations, public bodies, non-profit organizations, etc., appoint the members. Regardless of the rhetoric or effectiveness of the Council, many wonder if it can be sustainable after the Roh Administration.

The I Love Daegu Citizen Conference was established in 1999 by 136 organizations including NPOs, NGOs, religious organizations, press, and administrative bodies in Daegu. This citizen conference has been working on, among others, the "Fence Elimination campaign" as a major project and deserves appreciation especially for its early commitment and achievement. Recently, this seemingly huge organization has functioned in a fairly limited fashion in various decision-making processes. Instead it appears to have become oriented toward performing a symbolic role. Now may be the right time for the conference to redefine its identity as an active voluntary citizen group.

6.6 Closing

In this report, we reviewed how Daegu region has evolved in recent decades and what current issues exist in regard to spatial regeneration and local governance. At the outset, the physical development of the region has been outlined in the larger context of Korea's decentralization and national planning mechanism.

This is not purely a research paper but rather a case report, and the Daegu experience is perhaps not a completely successful case to learn from. Yet some insights and policy implications can be drawn from the examination and analysis.

First, the metropolitan Daegu region has experienced a great deal of inner-city transformation (perhaps, rather than urban regeneration) and suburban growth; the physical reshaping appears to be on-going process. The social and economic changes that cannot be decoupled from the spatial rearrangement in the process are also to be noted.

Second, one cannot say that Daegu's physical changes have been initiated or carried out by the local government alone. Market forces have been more likely taking the lead, not only in physical development but also in socio-economic changes, which has particularly been the case in recent years. A closer look would imply that Daegu is in need of real urban regeneration as the city matures further.

Third, the region has suffered from considerable economic difficulties for decades, especially compared to the Capital Region. Combined efforts by Daegu and its neighboring Gyeongbuk have been emphasized to restructure the regional economy. The major potential lies in the regions' rich intellectual resources. Not always welcomed by all people, the campaign for Daegu-Gyeongbuk economic integration, or the campaign for cooperation rather than competition, is a significant springboard for survival and the advance of the two jurisdictions.

Fourth, it is also to be emphasized that the whole process of urban regeneration could not be satisfactorily implemented without good local governance. Since the mid-1990s, a variety of non-governmental entities have taken part in managing local communities, which will be anything but dwindling in the years to come. A more decentralized and participatory governance in the 'booming' civil society of Daegu is strongly recommended.

Notes

[1] Some policy means already in place include the following:
(i) provision of sufficient urban infrastructure or capital facilities (general requirement)
(ii) impact fee for urban infrastructure (first introduced and levied nationally in 2006)
(iii)'development density management area' (not yet designated)

References

Daegu Metropolitan City (2005) *2030 Daegu City Master Plan* (in Korean)

Daegu Metropolitan City (2007) *International Knowledge Industry City Project (draft)* (in Korean)

Daegu-Gyeongbuk Development Institute (2006) *Tell Me True Daegu* (in Korean)

Kidokoro, T. (2006) *Sustainable Urban Regeneration and Local Governance in Regional Cities* (Presented in International Workshop on Sustainable Urban Regeneration and Local Governance in Regional Cities. January 17-18, 2006. Tokyo, Japan)

Kim, S.T. (2003) "Governance System for a Prestige City" in *Daegu Metropolitan City, Long-term Development Plan for Daegu* (in Korean)

Kim, W.B. (2005) *Korean Cities in Search of Identity* (Presented in International Conference: Revitalization and Globalization of Daegu-Gyeongbuk: Learning from Abroad. November, 2005. Daegu-Gyeongbuk Research Institute)

Korea Research Institute of Human Settlement (KRIHS) (2000) *The 4th Comprehensive National Territorial Plan in Korea (2000~ 2020)* (Research Monograph 2000-5) (in Korean)

Korea Research Institute of Human Settlements (KRIHS) (2005) *Daegu Metropolitan Urban Plan. A report to Daegu Metropolitan City & Gyeongbuk Province* (in Korean)

Lee, C.G. and Kwak, J.M. (2005) *Analysis of Regional Economy Impact by KTX* (Daegu-Gyeongbuk Research Institute Report) (in Korean)

Lee, S.C. (2000) *Measuring Acceptance of Regulatory Growth Management Policy: Korea's Green Belt Case* (Ph.D. Dissertation. University of Washington, U.S.A.)

Lee, S.C. (2007) "Recent Decentralization Challenges in Korea: Repertoire, Reality and Reshaping", *International Review of Public Administration*, 11(2)

Ministry of Construction and Transportation (2002) *Annual Report on National Land Use* (in Korean)

National Statistics Office (2005) *Population Prediction by Region* (in Korean)

National Statistics Office (2006) *GRDP Statistics by Region* (in Korean)

Onishi, T. (2007) *Practice and Perspective of the Urban Renaissance Policy in Japan* (The 2007 International Seminar on Urban Regeneration in Seoul)

Peters, G. (1996) *The Future of Governing: Four Emerging Model*. University of Kansas Press

Research Group for Economic Integration of Daegu-Gyeongbuk (2006) *The Plan for Economic Integration and Development of Daegu-Gyeongbuk* (in Korean)

Rhodes, R. (1996) "The New Governance: Governing without Government", *Political Studies*, XLIV

Ryu, B. (2004) "Daegu's Governance" in *Daegu Agenda 21* (in Korean)

Stoker, G. (2003) *Transforming Local Governance: From Thatcherism to New Labor*. Palgrave

Daegu Metropolitan City Government, www.daegu.go.kr (Accessed in January 2007)

Ministry of Construction and Transportation, www.moct.go.kr (Accessed in December 2006)

Ministry of Government Administration & Home Affairs, www.mogaha.go.kr (Accessed in June 2007)

Multi-functional Administrative City Construction Agency, www.macc.go.kr (Accessed in December 2006)

The City of Daegu, www.daegu.go.kr (Accessed in 2007)

7. Urban Form and Rail Transit Development in Dalian, China

Shengchuan Zhao and Zhiwei Yang

7.1 Introduction

Dalian, known as 'the Pearl of Northern China,' is located on the east coast of Eurasia and the southern tip of the Liaodong Peninsula. It stretches from 120°58' to 123°31' east longitude and 38°43' to 40°10' north latitude, with the Yellow Sea on the east, Bohai Sea on the west. Because of the superb geographical location, Dalian serves as a marine gateway to Beijing and vast northern China.

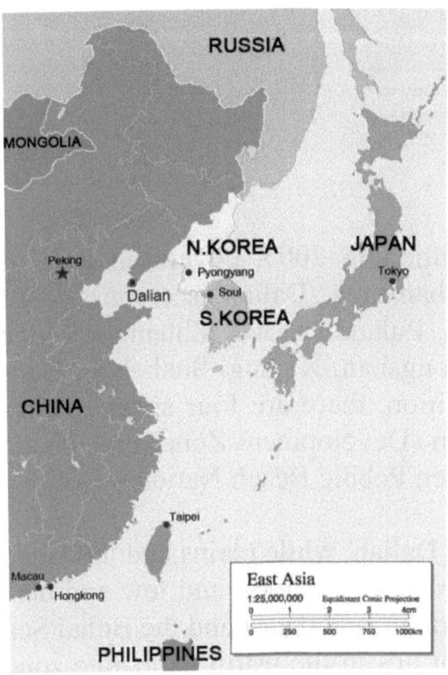

Fig. 7-1. Location of Dalian

Table 7-1. Profile of Dalian in 2005 (Source: Dalian Statistical Yearbook 2006)

Area (km^2)	12,573.85
Population	5,653,325
Population density (person/ha)	4.5 (Municipality) 132 (3 central wards) 11.6 (6 central wards)
Household size (person/hh)	2.9
Percentile of senior people	8.7 (2000)
GRP	215 billion RMB ($27.6 billion)
GRP per capita	38,070 RMB ($4,881)
Industries component (1st:2nd:3rd)	8.5:46.3:45.2
Trips by rail transit per capita	12.7
Trips by bus per capita	195.3
No. of cars per 1000 pop	56.1

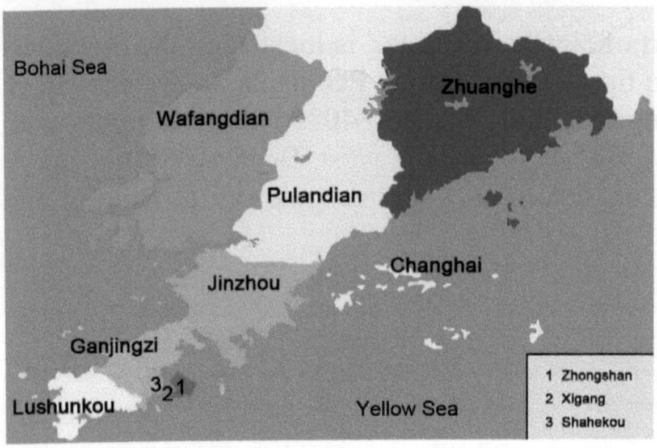

Fig. 7-2. Dalian Municipality

Dalian, with a population of 5.7 million in 2005, covers an area of 12,574 km^2, of which 2,415 km^2 is urban area. Dalian is comprised of three county-level cities (Wafangdian, Pulandian and Zhuanghe), one county (Changhai) and six districts (Zhongshan, Xigang, Shahekou, Ganjingzi, Lvshunkou and Jinzhou). In addition, there are four special zones under the direct administration of Dalian (Development Zone, Free Trade Zone, Hi-Tech Industrial Zone and Golden Pebble Beach National Holiday Resort).

An undulating topography is found in Dalian, while plains and lowlands are rarely seen. The terrain, high and broad in the north and low and narrow in the south, tilts to the Yellow Sea on the southeast and the Bohai Sea on the northwest from the center. Dalian lies in the warm temperate zone of the Northern Hemisphere, with the maritime feature of a warm temperate continental monsoon climate. With an average annual temperature of 10.5 °C, four seasons in Dalian are distinct, yet have neither extremely cold weather in winter, nor extremely hot weather in summer.

Dalian was renamed from Qingniwa in 1899 during the Russian occupation. In the Sino-Japanese War of 1894-1895 and the Japanese-Russian War, Dalian suffered great damage and then became a Russian and later a Japanese colony for nearly half a century. China resumed the government of Dalian in 1951, after which socialistic construction in the city began. From 1966 to 1976, Dalian experienced the Cultural Revolution, a 10-year period of confusion, when city construction stagnated. After 1976, Dalian entered the new period of socialistic modernization construction. Approved by the State Council, Dalian became one of the coastal 'opening-up' cities in 1984, and was designated as a city with a separate economic plan in 1985, enjoying provincial-level decision-making authority. Dalian has been experiencing rapid industrialization, urbanization and motorization since the 1990s. In 2003, the central government made an important decision, that is, to build Dalian into an international logistic hub in Northeastern Asia, and Dalian set out on a new journey of accelerating its internationalization.

7.2 Master Plans and their Impacts on Urban Development in Dalian

With more than 100 years' development, Dalian has become an important port, and a city of trade, industry and tourism in northeastern Asia. The framework of modern Dalian was shaped during the Russian Lease Period (1898-1904) and the Japanese Occupation Period (1905-1945). Dalian had become a world famous port city by the 1940s. The population of the city grew from the beginning to the middle of the 20th century from 40,000 to nearly 700,000. Since the foundation of the People's Republic of China, Dalian has seen rapid economic development and experienced the transition from planned economy to market economy, after which the city's current layout was finally formed.

7.2.1 The Russian Lease Period (1898-1904)

The basic urban form of Dalian was shaped during the Russian Lease Period, which lasted for seven years. In 1898, Russia took the lease of Lvshunkou and Dalian Area, making Lvshunkou as its military base and decided to establish Dalian as a major commercial port linking Europe and

Asia. The Russian authority designed the city and port master plan of Dalian in 1899 and started the city's construction process, aiming to make Dalian a port and commercial city.

Russian authority's urban planning for Dalian focused on transport and the city's layout. It adopted European classical formalism and the planning region reached 6.5 km2, which divided the city into three sections by functions: the administrative area, the European area and the Chinese area. In those years, population reached to 41,260.

Considering the geographic position of Dalian, Russian's planningfocused on the construction of harbor and railway that became the engine of the city's development in later years.

Russia's planning on Dalian demonstrated strong features of a colonial city. It has been proved that the land transport network planned by Russia couldn't meet the needs of the era of automobiles that came later. Nevertheless, the Master plan has a number of characteristics to be mentioned:

- reasonable arrangement of coastlines
- good integration of port and road network
- large areas for parks and public green spaces (16%)

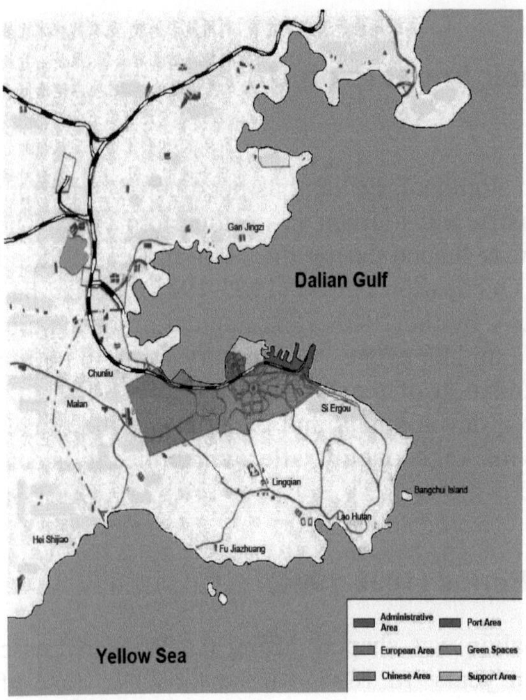

Fig. 7-3. Masterplan of Dalian city in 1901 (Source: Dalian Municipal People's Government (1999))

7.2.2 The Japanese Occupation Period (1905-1945)

Japan continued the construction of Dalian after occupying the Lvshunkou and Dalian area, based on the Russian urban planning. Japan decided to build Dalian into a 'free port.' In 1909, the Japanese authority compiled *The Planning of Dalian Urban Area*, in which regions of the city that had been accomplished by Russia remained unchanged, while modifications and adjustment were made to those unfinished during the Russia era.

The urban population of Dalian grew to 108,000 as the city's economy developed rapidly. The urban area was in short supply. The Japanese authority made The Scheme for Road and Street Expansion in 1919, which rendered the city's expansion toward the west. Thus, the size of Dalian's urban area expanded from 16.7 km^2 in 1919 and 35.6 km^2 in 1924 to45.7 km^2 in 1945. The first tram line was started in 1909, and three lines totaled 9.8 km at that time. The urban planning and construction during this period shaped the basic framework of Dalian, having far-reaching influence on the urban planning of Dalian in later years.

The scheme for the district planning of Dalian was made in 1934 to meet the demand of urban development. It was considered to be the first comprehensive urban planning of Dalian, which contained many prospective proposals. But the planning scheme was not implemented since Japan was distracted by the Second World War that broke out later.

Like Russia, Japanese planning of Dalian demonstrated strong features of establishing Dalian as a colonial city. In the Japanese mind, everything

Fig. 7-4. Master plan of Dalian city in 1934 (Source: Dalian Municipal People's Government 1999)

in the city served military and economic purposes. All the regions alongside the Dalian Gulf were occupied by wharfs and factories instead of living areas. The Japanese authority brushed aside the city's ecological environment, and destroyed a number of green lands. A number of low-lying lands caused by random sea reclamation made the city vulnerable to flooding. And the release of untreated wastewater brought pollution to the Dalian Gulf. And those problems remain difficult to solve today.

7.2.3 Centrally-planned Economy Period (1945-1978)

Japanese surrendered in 1945, and a Russian presence was felt until 1955. The urban planning activities stopped during 1945-1955. After the foundation of the People's Republic of China, the economy of Dalian was improved under the guidance of a three-year economic recovery plan and a period of intensive construction guided by the first Five-Year Plan Program. Starting from 1957, Dalian began to draw up the latest urban planning scheme. Designed in the year of 1958, the new scheme was called the "1958 Plan". It was the first city master plan for Dalian since the foundation of People's Republic of China. The accomplished urban region at that

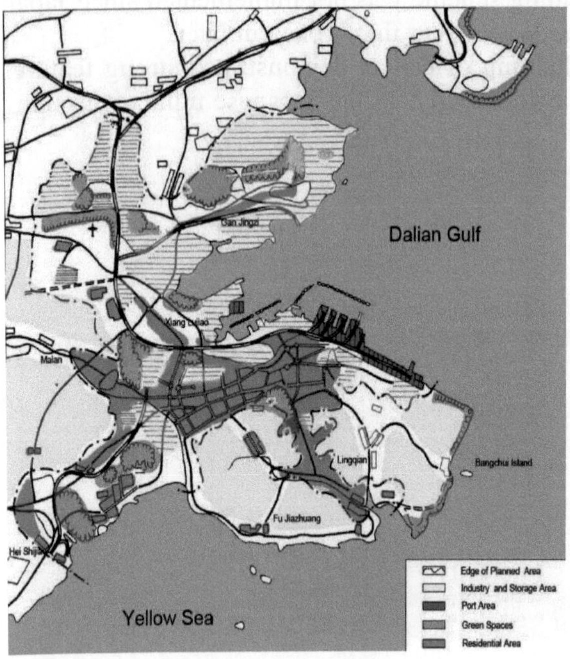

Fig. 7-5. Master plan of Dalian city in 1958 (Source: Dalian Municipal People's Government (1999))

time covered 70 km^2, with a population of 800,000. The city's economy was dominated by heavy industries focusing on the machinery and chemical, as well as, transportation industries focusing on harbor and railway construction.

The guideline of the "1958 Plan" was that "the city should serve the needs of industry, agriculture and national defense, as well as the people." Accordingly, Dalian was supposed to be constructed as an industrial city economically led by machinery and chemical industries, with the urban area covering 102 km2 and the population of 1,300,000. The lands for industrial use took the priority in the planning which set up 10 industrial zones in the city. Meanwhile, the planning also included the arrangements regarding the road system, municipal facilities, green land, medical and health care, commercial service, resident zone, reform of old districts, etc.

Generally speaking, the "1958 plan" was reasonable regarding some of the principle items such as the scale of the city, city's layout, use of the land, and functional divisions. But as the impacts of the Great Leap Forward policies and Russian urban planning concepts, excessive investment in industry caused the blind development of the city in later years. Moreover, the plan makers had not noticed the role of the port, which should be the engine of Dalian's economy.

7.2.4 Economic Reform Period (1978-1990)

From 1966 to 1976, Dalian experienced the Chinese Cultural Revolution, which ceased urban planning for more than 10 years. Because of economic reform and open-door policy, Dalian entered the new period of socialist modernization construction after 1978.

Dalian updated an overall Masterurban in 1980, which was called the "1980 Plan" for short. Granted by the State Council in 1985, the "1980 Plan" focused on the construction of four center areas in the city (the Master plan for Lvshunkou was designed separately), and identified Dalian as a port, industrial and tourist city, with a territory of 118 km2 and a population of 1,400,000.

The "1980 Plan" attached great importance to harbor development and proposed to construct the Da Yaowan Harbor and developed the small and medium-sized ports. It also adjusted the layout of lands for industrial use, arranged 13 industrial zones and provided that there should be no more industrial projects in the old downtown area. It also planned to build up the

Dalian Economic and Technical Development Zone in Maqiaozi village of Jinzhou county, and to attract more foreign investment, as well as develop knowledge-intensive industries. The plan also proposed to develop

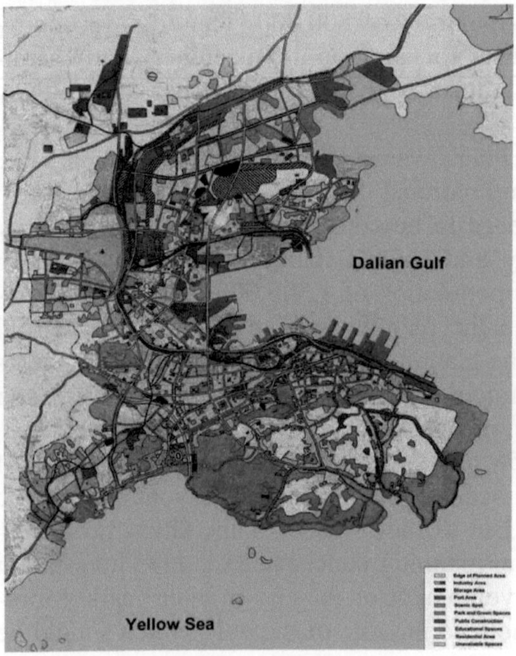

Fig. 7-6. Masterplan of Dalian city in 1980 (Source: Dalian Municipal People's Government (1999))

the tourist zone, strengthen environmental protection, and step up efforts on the construction of small towns. Thus, the construction plan for the short term was worked out. The road network was also expanded at that time, and Shenyang-Dalian Expressway was completed.

The "1980 Plan" came in time and was considered as a good planning scheme that served as the guide of the city's construction. The economy of Dalian developed rapidly in the following 10 years and the city's economic status in China was enhanced consistently. But there were still some shortcomings in the plan. It underestimated the speed of the economic growth of Dalian, for the index of the city scale in 1989 already exceeded the standard set for 2000.

7.2.5 Rapid Economic Growth Period (1990-1999)

Since the goal of the "1980 Plan" had been exceeded ahead of schedule, the Dalian government worked out "The Adjustments of Urban planning of Dalian" in 1990, which was called the "1990 Plan." for short.

No modifications were made in terms of the nature of the city in the"1990 Plan", but the planning concept of "not seeking for the biggest

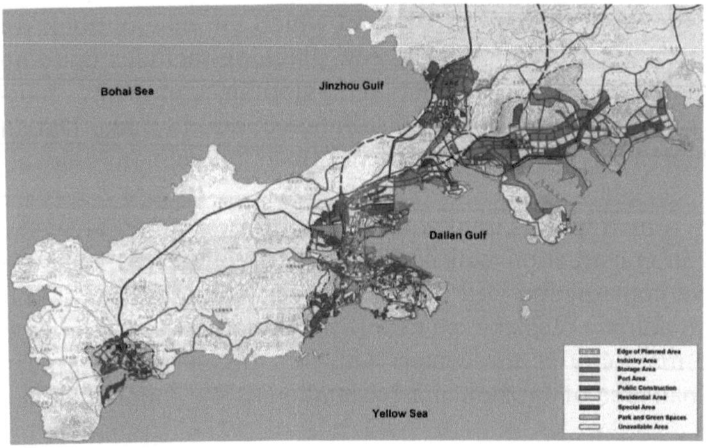

Fig. 7-7. Master plan of Dalian city in 1990 (Source: Dalian Municipal People's Government (1999))

but the best" was introduced by former mayor Bo Xilai, and the long-term goal for the city's development was raised in the plan; that is, "to build Dalian into an international city." Modifications were made defining the region that was included in the planning, the prospect and structure of the city's development, the system of the city and towns, and the city's infrastructures, etc. In the plan, the concept of Dalian as a city was first set and it proposed to develop cluster-type structure as the basic structure of the city. It also defined the functions of each cluster in order to benefit the planning and development of each district.

The proposal of building new urban areas was added in the plan. It proposed to develop new ports and build up new urban areas in the Economic and Technological Development Zone, Dayaowan Harbor area, Wanli, and Dong Jiagou. The strategy of the city's development was included into the planning text for the first time, which allowed the major infrastructure programs to be implemented in the city in advance. The "1990 Plan" thus updated the "1980 Plan." Both served as the guidelines of construction of Dalian in the period of Chinese economic reforms and opening-up, and laid out a solid foundation for Dalian to step forward toward an international modern city.

7.2.6 Transition towards a More Market-oriented Economy (2000-present)

Formulation of the "2000 Plan" was started in 1996, with an outline proposed in 1998 and its first version completed in 2000. The "2000 Plan" was approved by the State Council in 2004, in which the new outlook for development of Dalian was proposed. The new planning includes building Dalian into a regional center for international shipping, commerce, tourism, finance and information in the East Pacific area, and to make Dalian an international tourist city that have radiative effects on economic development of Northeast Asia.

It plans to form a metropolis circle within 2,415 km^2 of urban area; the land used for urban construction will be constrained within 450 km^2; the land used for town construction will be constrained within 180 km^2. And the whole city will adopt a cluster-type development pattern that holds the urban area of the main city as the center, with new urban area as a sub-center, and the urban area of Jinzhou and Lvshunkou as the supplementary sections.

It plans to strengthen efforts to build a new international airport, to construct major roadways and railway terminals, and to strengthen radiative effects of Dalian as a modern center city and as an important transport terminal in Northeast China. The harbors of Dalian and Dayaowan will be the main seaports in Dalian. Combining with the newly constructed airport, the Harbin-Dalian electrical railway, and the expressways connecting Shenyang-Dalian, Heilongjiang-Dalian and Dandong-Dalian, together make up Dalian's three-dimensional transportation system.

This system will focus on the inter-city transportation system mainly supported by Mass Rapid Transit (MRT), strengthening the construction of urban rail transit system, developing the intelligent transportation system and the high-tech transport management system. At the end of 2004, the

Fig. 7-8. Development mode of Dalian city in "2000 Plan" (from Dalian Municipal People's Government 2000)

MRT No.3 route started its full services.

Currently, Dalian is facing unprecedented challenges and opportunities. Dalian needs to adjust itself to the new environment and the rules of competition between cities under the new context. Various aspects need to be considered including industry, culture, space, operation and innovation, in order to make the new strategic master plan of Dalian more reasonable.

7.3 Rail Transit Development in Dalian

Rail transit is not a new concept for city of Dalian, which has been developed with tram line construction. The rail transit system in Dalian consists of municipally operated streetcars, trams, the LRT and MRT. In 1908, the tram line from Dalian Seaport to the former Zoo began to be constructed, and was finished with a length of 2.45 km after one year. This tram line was the first one in Dalian, even in northeast region of China, and Dalian entered an era of modern public transportation. After that, construction of tram lines was carried on for decades. By the end of 1939, the tram line in Dalian grew into a system composed of 11 lines, with total length of 32.7 km, which became the city's dominant transportation mode.

After the unconditional surrender of Japan on August of 1945, the Dalian municipal government took over the tram company, and carved out an economic recovery. The efforts to improve the comprehensive development of the tram proved effective. In 1960, the first municipally operated streetcar line from Dalian Railway Station to Quanyong Street with a length of 4.5 km was developed. In the following period, much updating was done. But because the public transit policy favored buses, many tram lines were taken over by motorbus. In the late 1970s, only three tram lines and two municipally operated streetcar lines remained.

In 1980s, the city government began research on the strategy of urban-passenger transport development, and decided to give development priority to public transportation, especially rail transit. The rail transit master plan was developed in 1984. The first stage of feasibility study of this rail transit project was approved by the Central Government in 1993. A number of rail transit routes were proposed in the transportation master planfunded by Japan International Cooperation agency (JICA) in 1996. The city government formulated a rail transit master plan in 1999. This plan included five routes with a total length of 123 km. These routes are either providing transportation services for new residential areas and industrial parks, or connecting with tourist spots.

To meet the need of Dalian's urban area expansion, MRT No. 3 route

Fig. 7-9. LRT (left) and MRT (Route No.3) (right) in Dalian

Table 7-2. Details of Dalian MRT system (Route No.3)

Item	Details
Project targets	Access to industrial park & tourist spot
Length	49.15 km, stations, 11 stations (7 elevated), station spacing 4,915 m
LOS	Average speed: 65 km/h, head: 10 min. (peak), 20 min.(off peak) Starting fare: 1 yuan ($0.13), max 8 yuan ($1)
System	Gauge: 1435 mm Radius: 300 m Slope: 2% 4 vehicles/train With driver and no platform door
Construction cost	Total: 2.34 billion yuan Average: 47.6 million yuan ($6 million/km)
Number of passengers	33,000 person/day 671 person/km

was developed and started its full services in 2004 (Fig. 7-9). The subway route proposal was included in the recent Master plan. In recent years, rail transit system has played an even more important role in new town development, port and tourist spot developments, which must lead to a new city configuration.

7.4 Interactive Impacts of Rail Transit and Urban Form

To meet the living demand of citizens of Dalian, a suitable urban form must be introduced, rather than traffic system. But the traffic system will certainly bring about the feedback to urban form, which could properly adjust urban structure and the density of local land development. Transit oriented development is a resource-conserving, environment-friendly, land-intensive, and growth-smart development pattern, and has been a strategic choice of urban development in Dalian. From the development of rail transit system in Dalian, we have learned some useful details of interactive impacts of rail transit and urban form.

7.4.1 Rail Transit Impact on Urban Form

As one of the pioneers of rail transit development in China, Dalian has adopted a strategy of rail transit oriented development in its early years, determined the original urban form, before the foundation of the People's Republic of China. With land distribution and rail transit system construction, the administrative and living districts developed along the tram line, bringing convenience to the city dweller.

Since the 1950s, due to the centrally-planned economy, the self-contained community was formed. Within this community, motorized transport was not necessary. Employers provided pick-up bus services for their employees who live outside of the community. Due to a rationing system and state-owned land system, the property market did not exist until the 1990s. Thus the transit system seems to have had little effect on urban form during the 1950s-1990s.

After the 1990s, along with transition from planned economy to market economy, more attention was paid to the new town development in Dalian, which was guided by the transit connection from the early planning stages. The commencement of LRT services has reduced the distance between

Table 7-3. Interactive impacts of rail transit and urban form

Transit impacts on urban form and land use	Downtown impacts	Little (during 1950-1990) and might be significant after 1990s due to the introduction of property market
	Impacts in suburban areas	Decentralization (relocation) of population & employment; edge cities or tourism development or connection with downtown
	Impacts on property values & rents	Weak at downtown; strong in suburb
Urban form & land use impacts on transit demand	Density & travel behavior	Depending on traveler's socio-economic characteristics (traveler is likely to drive once he/she owns a car as the car is a symbol of social status)
	Transit usage by proximity to stations	yes

the old and new towns, and thus most residents in old town of Dalian can save 50% travel time to reach most of the employment centers in the New Town.

As an economic center of north China, Dalian has an increasing population due to the fast development of the economy, and the antinomy between the shortage of land in the city center and rising demand is bringing more and more trouble. The development of rail transit releases the pressure in the circumjacent areas of the city, which helps to release the antinomy. There have been many rising commercial areas along the tram line in recent years. With the extension of the rail transit, some modern elements have appeared on the horizon. The rail transit in Dalian has even become an attractive part of the city landscape.

7.4.2 Urban Form Impact on Rail Transit

After decades of construction and operation, the urban form shaped by the rail transit system and other related public transportation systems have proven to function efficiently. With the development of urban form, some certain contributions to the sustainable development of rail transit have been made.

Together with the development of new towns in Dalian, industrial parks and suburban offices generate a large number of transit trips to and from work on weekdays. The only roadway connecting the old town and the new town in Dalian has experienced dramatic traffic congestions. After the MRT system became a connector between the two large settlements of Dalian, it has been possible to reduce auto dependency in these areas.

With the exploitation of tourism resources, the rail transit system is bringing convenience to tourists and tourist spots in Dalian attracting a large number of transit trips at weekends and holidays.

7.5 Towards a Sustainable Dalian

Under a high economic growth, Dalian continues to be faced with rapid urbanization and motorization. The City government has put its public transportation system as the priority in the city development. Upon the several decades of development of rail transit development, Dalian has achieved the following experiences:

- Dalian has a stronger authority to increase mobility. It has a single city (equivalent of a metropolitan region in Japan) government rather than being separated into a number of local administration units and also supportive policies for public transportation by the Central Government.

- Dalian has fewer barriers to new development due to a state-owned land use system. By transferring urban land use rights to developers, the City government can obtain sufficient funding to grant former city development projects.
- As the transit system is operated by a state-owned company, the City government supports the capital funding as well as operation costs of transit system.
- Dalian has been a public transportation oriented city having a long history of transit systems, and it is relatively easy to reach concessions in building new transit systems.

However, there are a number of issues to be solved for Dalian:

- A society 'ruled by laws' to enforce the Master plan
- Continuity of strategic urban development objectives
- Integration of urban planning and land administration
- Integration of different transportation authorities divided by different modes
- Public involvement by base-level governments
- Reduction of disparities by household registration (*Hukou*)

These issues will be challenges not only for Dalian but also for the nation. Promotion of unrestricted mobility confronts China, as rapid industrialization, urbanization and motorization moves forward. Strong regional vision of a desired urban form combined with political leadership anticipating the long-term benefits of a transit investment, is one of the prerequisites for the success of city development. It is projected that the total length of rail-based transit systems will exceed 1700 km in more than 15 cities by 2015. Thus the struggle continues for a sustainable Dalian, aiming at a resource-conserving and environment-friendly society.

References

Dalian Municipal People's Government (1999) *Dalian City Planning 100 Years* (edited by Liu, C.) Dalian Maritime University Publishing House

Dalian Municipal People's Government (2000) *Dalian City Masterplan 2000-2020*

Fouracre, P. and Dunkerley, C. (2003) "Mass rapid transit systems for cities in the developing world", *Transport Reviews*, 23(3)

Gakenheimer, R. (1999) "Urban mobility in the developing world", *Transportation Research A*, 33

Newman, P.W.G., and Kenworthy, J.R. (1996) "The land use-transport connection: An overview", *Land Use Policy*, 13(1)

Ryan, S., and Throgmorton, J.A. (2003) "Sustainable transportation and land development on the periphery: A case study of Freiburg, Germany and Chula Vista, California", *Transportation Research D*, 8

Transportation Research Board (1995). "An evaluation of the relationships between transit and urban form". *TCRP digest*

Zhao, S. (2006) "A comparison of transit developments in Hiroshima and Dalian: Implications for Chinese cities" (Proceeding of the 5th International Conferences on Traffic and Transportation Studies (ICTTS), Xi'an, August 2006)

Part III

JAPAN

8. Urban Regeneration and the Shift of Planning Approaches: The Case of Japanese Regional Cities

Tetsuo Kidokoro

8.1 Challenges Facing Regional Cities

8.1.1 Decreasing and Aging Population

Japan's population started to decrease in 2005. Large cities with populations exceeding 1 million people are forecasted to continue seeing population increases until 2015, while the populations of small to medium-sized cities have already started to decline. Thus, the competition for growth among small to medium-sized cities is particularly harsh. Another demographic issue to consider is aging. Japan is the world's fastest aging society and it is predicted that nearly one-third of the total population will be over 65-years old in 2030. It is surely a big challenge to find ways to maintain social vitality in an aging society. Rural areas surrounding regional cities are aging much more quickly than large metropolitan areas such as Tokyo. Thus, regional cities that serve as central cities of fast aging rural areas play important roles as driving forces of regional economic development.

8.1.2 Impacts of Deregulation Policies

Among the deregulation policies started since the 1980s, the measure that has most impacted urban spatial structure is the Retail Shop Location Law of 1998, which replaced the Retail Shop Coordination Law. Under the old law, when large-scale shopping centers (SCs) were developed, closed-door coordination took place between SCs and existing retailers regarding the location, size and operation time in order to protect existing retailers. This old law became a source of trade conflict between Japan and the US, since it was regarded as a non-transparent barrier preventing new retailers from entering the market and inhibiting efficiency in the retail sector. The new law changed the SC regulations, aiming to be more rule-based and regulate SC location and size soley based on neighborhood environmental impacts such as traffic congestion, without considering how SC development would affect existing commercial activities. Since then, a rush to develop new regional-scale SCs in urban fringe areas started throughout the nation to fill the vacuum in the market. And, of course, the city-center commercial functions have been severely damaged in regional core cities.

As for deregulation in urban planning regulations, after the burst of the bubble economy in 1990, various deregulation measures have been undertaken to vitalize urban development and reverse the trend of decreasing land prices, so that the bad debts problems of banks can be solved (loans to firms are normally secured through real estate collateral). Deregulation in urban planning allows for additional FAR bonuses. One of the deregulation measures is the Special Law on Urban Renaissance, enacted in 2002. Under this law, prefectures can designate special districts where deregulation measures to increase FAR can be applied.

Banking, information, research and development functions as well as oversees functions concentrate in the Tokyo metropolitan region, which acts as a global hub in the age of globalization. Thus, most urban development projects enjoying deregulation measures concentrate in the city center of Tokyo to accommodate new demand in response to globalization and informatization of the economy. But, very few cases are observed in regional cities, simply because they do not have enough demand to utilize the bonus FAR.

8.1.3 Declining City Centers of Regional Cities

Globalization affects cities in different ways, largely reflecting the position in the urban hierarchy within the nation and the world. Creative functions such as central management, R&D, IT industry, etc. tend to concentrate increasingly in global gateway cities, in particular Tokyo in the case of Japan. On the other hand, many regional cities have lost jobs in the manufacturing sector due to harsh competition on the location of factories with other countries, particularly Southeast Asia and China. Furthermore, suburbanization and increasing dependence on automobiles as well as the development of large suburban shopping centers have directly undermined the position of the city centers. Ironically, relocation of public buildings such as local government offices, hospitals, universities, cultural centers baseball parks, etc., to suburbs further aggravated the problematic situation of city centers in regional cities.

8.1.4 The Beginning of Regional Considerations?

Faced with the severe decline of city centers, the City Planning Law was amended in 2006. The main aim of the amendment was to revitalize declining city centers, and it imposes regulations on the location of SCs with floor areas over 10,000 m2 and public buildings. From 1998 until 2006, the locations of SCs could be decided rather freely depending on the availability of land and accessibility by cars largely in urban fringe areas. Under the amended City Planning Law of 2006, the locations of newly developed SCs are limited to three land use zones: the commercial zone, the neighborhood commercial zone, and the semi-industrial zone designated by the statutory City Planning Map. Public buildings were formerly exempt from development permission and thus could be located anywhere. Many public buildings indeed moved from city centers to suburban areas or were developed in suburban areas in the search for available land. This has been identified as a factor behind the decline of city centers.

However, it cannot be said that the revitalization of city centers is on track. The current tendency of national government policy is to support winning horses in order to create driving forces for national economic development. Under this neo-liberal approach, competition among cities is growing increasingly intense. If the term ‘sustainable urban regeneration’ is defined as ‘enhancing the attractiveness of cities in response to changing external social and economic conditions,’ it entails cultivating and developing local resources through expanding social, economic and knowledge networks. This chapter examines current efforts in Japanese regional cities and discusses their issues from the viewpoint of the formation of local governance and planning approaches.

8.2 Overview of Urban Planning Systems in Japan

Japan’s current urban planning system was established by the City Planning Law of 1968, which replaced the old City Planning Law of 1919. Though the law has been amended from time to time since then, the basic structure remains same: the national government sets a framework which is uniformly applied throughout the whole nation and local governments (either municipalities or prefectures) formulate and implement the plans. There are two tiers of local governments (47 prefectures and about 1,800 municipalities). Prefectures have elected governors and elected prefecture assemblies. Municipalities have elected mayors and elected municipal councils; thus, politically, municipalities are independent from prefectures. The duty of prefectures is seen as dealing with issues that affect areas beyond municipal boundaries in the urban planning system.

The basic structure of the urban planning system is shown in Fig. 8-1. The prefecture designates the City Planning Area (CPA) where the City Planning Law is applied. CPA may cover several municipalities or can be applied to a single municipality. The coverage of CPA should ideally cover actual metropolitan areas, but often is decided based on political considerations among neighboring municipalities. Indeed, most CPAs are designated in a fragmented way and do not cover whole metropolitan areas. The prefecture formulates the City Planning Area Master Plan (CPA-MP). CPA-MP's most important function is to designate the Urbanization Promotion Area (UPA) and the Urbanization Control Area (UCA), while municipalities are responsible for the designation of land use zones (12

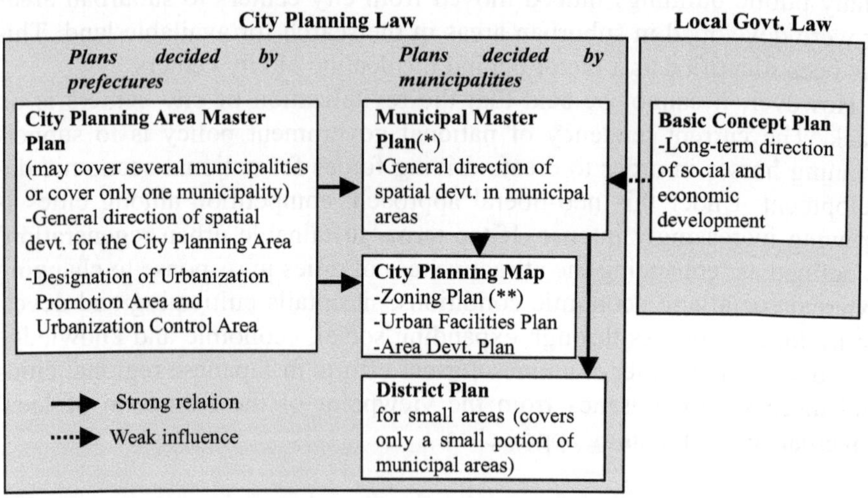

(*) The Municipal Master Plan should be in accordance with the City Planning Area Master Plan and the Basic Concept Plan. When decided by the municipality, the Municipal Master Plan as well as the Zoning Plan require the prefecture's consent.
(**) Zoning Plans for municipalities in the Tokyo, Osaka and Nagoya metropolitan areas are decided by the prefecture.

Fig. 8-1. Urban planning system in Japan

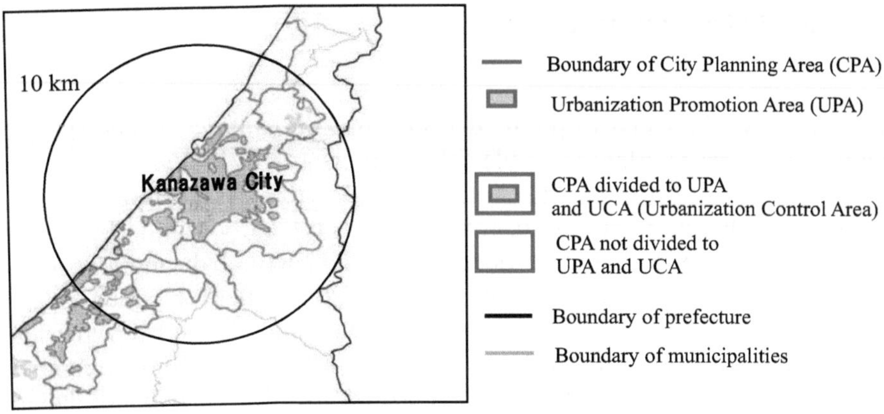

Fig. 8-2. City planning areas in Kanazawa city and its environs

categories), though consent from the prefecture is required for metropolitan-scale considerations.

If properly applied, UCA designation is a quite strong tool for preventing sprawl. In UCAs, urban developments are not allowed. CPAs covering cities with populations of 100,000 or more are basically to be divided into UPA and UCA in order to avoid sprawl in urban fringe areas under the guideline of the Ministry of Land, Infrastructure and Transportation in charge of urban planning. If this designation is to work effectively, CPAs should cover whole metropolitan areas and the division of land into UPA and UCA should be applied to all CPAs. But, due to a loophole in the system, unfortunately this is not the case. CPAs of smaller cities surrounding central cities are not necessarily required to be divided into UPA and UCA. Thus, suburban municipalities quite often oppose being divided into UPA and UCA or being included in the CPAs of central cities so that they can retain the possibility for urban development. The legitimacy of the division of land into UPA and UCA is therefore significantly undermined (Fig. 8-2). Moreover, urban developments are principally allowed in areas outside UCA as the landowner's right of development. As a result, urban developments are not allowed in the fringe areas designated as UCA in central cities, but conversely could be allowed in neighboring small cities. The amendment of the City Planning Law in 2006 was made partly in response to this problematic situation, though it only regulates the location of large-scale shopping centers. The problems regarding the location of residential developments remains unsolved.

8.3 Local Governance

Fig. 8-3 shows the styles of public decision-making, analyzed according to two planning issues: planning style (plan-led or project-led) and the extent of involvement of different stakeholders. Among the styles of public decision-making shown in the figure, corporatism is typically seen in European countries. This style of public decision-making occurs through negotiations among a few influential bodies such as government, businesses and labor unions. Pluralism seems rather characteristic of North America, where the legitimacy of public voting and the allocation of resource through market systems are widely supported. On the other hand, Japan, similar to China and Korea, has had a long tradition of public decision-making under established bureaucracy with strong hierarchy from the national government through the prefectures to municipalities, as seen in the hierarchical structure of the urban planning system.

However, a shift in public decision-making styles from government to governance is widely observed throughout the world. In Japan, too, the shift from government to governance is seen in response to the transition from pro-economic regional development in the past to sustainable development seeking balance among economic, environmental and social dimensions. Generally speaking, following approaches toward governance are observed in Japanese regional cities (Fig. 8-3): the corporatist plan-led approach, the market-oriented project-led approach, the ad-hoc citizen

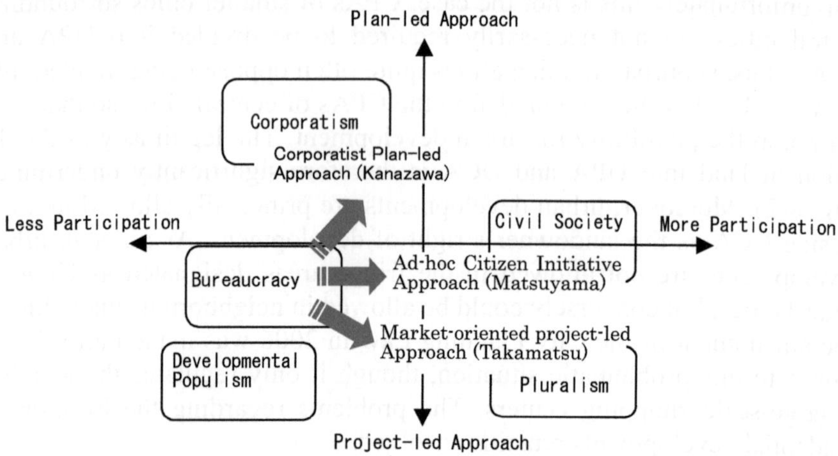

Fig. 8-3. Conceptual framework for governance style and planning approach regarding urban regeneration

Fig. 8-4. Location of case study cities

initiative approach.

The corporatist plan-led approach implements rather consistent urban regeneration policy through strengthening the partnerships among the mayor, bureaucrats, local politicians and the private sector leaders who have an interest in urban regeneration. The essence of the market-oriented project-led approach is the promotion of urban regeneration through deregulation measures to facilitate private investment in symbolic urban development projects that are expected to fuel revitalization. The ad-hoc citizen-initiative approach represents bottom-up attempts to regenerate cities through various private-sector associations and citizen groups.

The three cases (Kanazawa city, Takamatsu city and Matsuyama city) examined in this chapter seem to represent the three approaches respectively, and make it possible to grasp the current trends in the formation of governance and issues facing Japanese regional cities (Fig. 8-4).

8.4 Case of Kanazawa City

Kanazawa city (population of 456,000 in 2005) is the capital city of Ishikawa prefecture. The Hokuriku Shinkansen (high-speed rail) is scheduled to open in 2014, which will connect Kanazawa and Tokyo in 2.5 hours. The case of Kanazawa city represents the corporatist plan-led approach, in the sense that it tries to promote urban regeneration through cooperation between the city and city center business leaders and strengthening urban planning city ordinances. It is also noted that the corporatistic relation between the city and city center business leaders has been strengthened through partnership-based urban regeneration programs.

8.4.1 Urban Development Projects under the Initiative of Local Government

Kanazawa city was not damaged during World War II, and thus the urban structure of the historic town center remained untouched. As for new developments in suburbs after the war, almost 40% in the Urbanization Promotion Area (UPA) have been developed in a planned way through land readjustment projects, a very high rate among Japanese cities. Suburban agricultural land is mostly owned in rather small parcels (typical size: about 1 ha) by farmers (former tenants) who were allocated land through agricultural land reforms implemented after World War II. Under these conditions, land adjustment projects have played a major role for the suburban development in Japanese cities in the post-war period. Under land readjustment projects, land parcels owned by farmers are pooled and some 30-40% of the land is contributed for public facilities and sale to finance infrastructure development and the rest of the land is subdivided for residential plots. The subdivided land plots are reassigned to the original landowners so that they can retain the capital gain from the projects.

In Kanazawa, large-scale urban development projects have been promoted under the strong initiative of both the city government and the prefecture government, which was common in other regional cities, too. Indeed, by far the largest is the sub-center zone project, though most land readjustment projects were conducted for the purpose of suburban residential developments. In this project, the corridor connecting the Kanazawa train station with the Kanazawa port was designated as the sub-center zone. Half of the zone is developed by Kanazawa city and the other half by Ishikawa prefecture, including the development of the Prefecture Government Office Complex (Fig. 8-5). The concept of this sub-center develop-

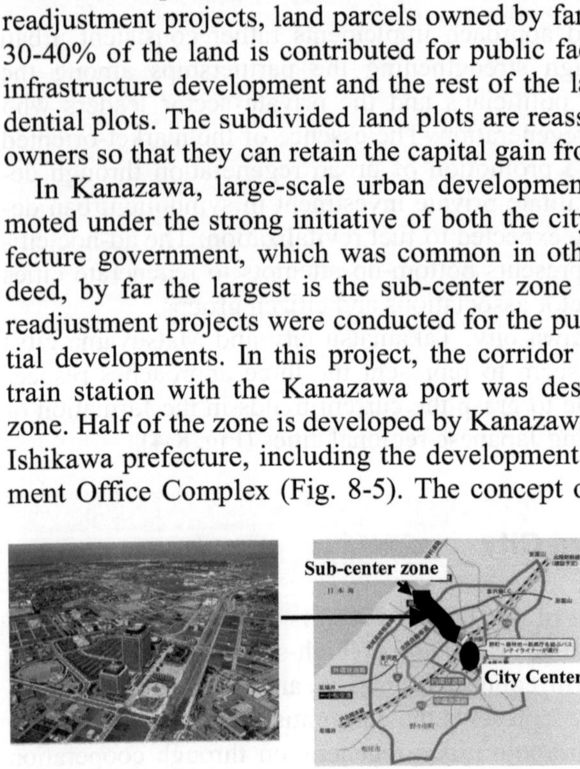

Fig. 8-5. Sub-center Zone Development, Kanazawa (Source: Kanazawa City)

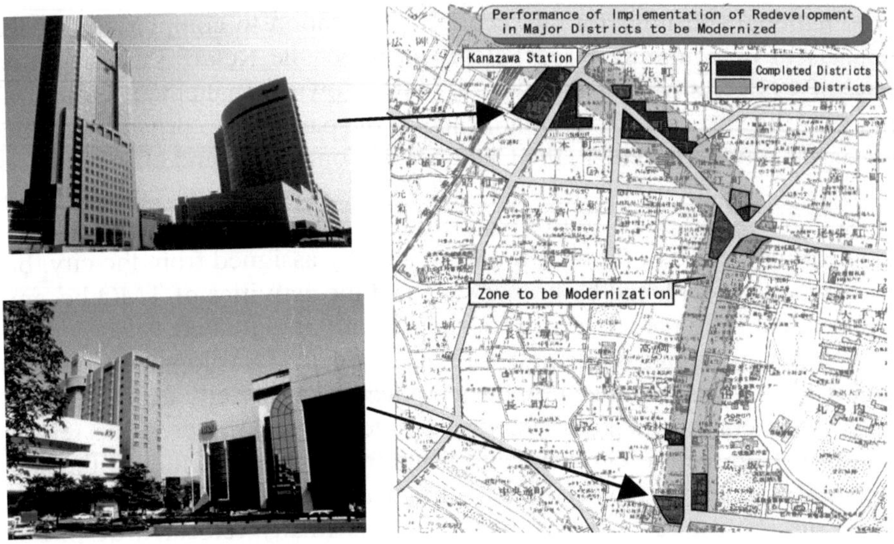

Fig. 8-6. Urban redevelopment in the central development corridor, Kanazawa (Source: Endo (2006))

ment project was first proposed in the Kanazawa Basic Concept Plan in 1970 and the project began in the early 1980s.

The corridor extending from the Kanazawa station to the city center, which is an extension of sub-center zone, was designated as the central development corridor and steady urban redevelopment projects have been initiated by Kanazawa city since the early 1980s (Fig. 8-6). Together with the sub-center zone development area, this development corridor from the city center through the train station to the port is intended to accommodate the new office demand expected in response to the opening of theHokuriku Shinkansen (high-speed train line).

However, the demand for new office development has been much weaker than expected so far, because under the trends of globalization and informatization firms are trying to integrate central management functions in Tokyo, and thus office demand in regional cities is weak. This challenges the the legitimacy of urban development initiated by local government.

8.4.2 City Center Corporatism: City Center Revitalization Projects

Kanazawa city formulated the City Center Revitalization Plan in 1998, based on the City Center Revitalization Law of 1998. This law strengthened national subsidies for city center revitalization to compensate for the impacts of deregulating SC development under the Retail Shop Location Law of 1998. As part of the plan, a TMO (Town management Organization) for the city center was established as the partnership organization between the city government, the Chamber of Commerce, and city-center retailers' association (Table 8-1). TMO's tasks were: management consultancy, marketing and organization of events, management of common parking, etc. The TMO staffs were initially assigned from the city, but they were later replaced by proper staff and the activities of TMO became independent from the city and were significantly invigorated.

A voluntary organization called the City Center Development Association (CCDA) was organized by business leaders in the city center, e.g. retailers' associations, department stores, hotels, etc., to oppose the relocation of the Prefecture Government Office from the city center to the sub-center for the fear of the decrease of commercial demand in the city center. After the Prefecture Government Office was finally relocated despite of the opposition of city center business leaders, the TMO took a role as the secretariat of the CCDA, and the organization changed its role to serving as a coordinating body among various stakeholders, working in collaboration with the TMO for city center revitalization programs.

This new collaboration between the TMO and CCDA became a new engine for regenerating the city center. A remarkable result of this collaboration is an infill-type commercial development project called Prego. The 1,700 m^2 plot in the midst of the city center had been vacant since 1982. This area was developed by the TMO in collaboration with the city, the prefecture, the national government as well as the CCDA (50.0% of the budget from national government subsidy, 24.3% from the prefecture, 24.8% from the city, and 0.9% from the TMO). As an organization representing city-center businesses, the CCDA took initiative in collaboration with the TMO to select the tenants to improve the attractiveness of the city center as a whole.

Table 8-1. Shareholders of the Kanazawa Town Management Organization (Source: Kanazawa Town Management Organization)

Shareholder	**No.**	**Amount (000yen)**	**%**
City	1	23,000	50.0
Chamber of Commerce and Industry	1	2,100	4.6
Retailers' Associations, Retailers	21	15,250	33.2
Large Retail Stores	4	3,630	7.9
Banks	2	2,000	4.4
Total	29	46,000	100.0

As illustrated above, through the formulation and implementation of city center revitalization projects, we can observe the process of strengthening corporatist relationships among a few key stakeholders (local governments and the city center business leaders) in the city center.

8.4.3 Setting the Regulatory Framework under City Center Corporatism

In Kanazawa, many regulatory city ordinances dealing with spatial development have been enacted recently. Among them, the most important is the Shopping Center Location Regulation City Ordinance of 2002. At the national level, the Retail Shop Location Law of 1998 deregulated the location of Shopping Centers (SCs), which promoted the development of many new suburban SCs and became one of the major causes of the decline of city centers. The mayor of Kanazawa city, a former city government bureaucrat, promoted the city ordinance to limit the location of SCs in Kanazawa city to preserve the attractiveness of the city center. This ordinance was also strongly support by the Chamber of Commerce and Industry of Kanazawa, whose chairman was the owner of the local department store located in the city center.

Furthermore, under the leadership of the mayor, a series of conservation regulations for the historic town center areas have also been enacted, including height limitations to preserve low-rise residential areas and restrict the construction of high-rise apartment buildings; townscape preservation ordinance; city waterway preservation ordinance; nighttime townscape ordinance; and an advertisement ordinance.

In the case of Kanazawa, urban development projects, which have been steadily implemented since the early 1980s, have been conducted under the strong initiative of the local governments (both the city government and the prefecture government). The aim of urban development was developmental and the style of implementation was project-based. The key promoter was governmental bureaucracy. This style of urban development has prevailed in post-war Japan both at the local and national levels.

Yet, since the late 1990s, faced with the decline of the city center, an arena for collaboration between the local governments and city-center business leaders has formed in Kanazawa. Based on the cooperation among the city and city center business leaders, a regulative framework in the form of a series of city ordinances has been strengthened. In this sense, it can be said that a new style of local governance and planning approach, that is, a corporatist plan-led approach, is observed in Kanazawa city.

8.5 Case of Takamatsu City

Takamatsu city (population of 338,000 in 2005), capital of Kagawa prefecture, represents the market-oriented project-based approach. It tries to promote urban regeneration through the attraction of private-sector investment by deregulating urban planning and conducting symbolic urban regeneration projects. It should also be noted that the initiative of urban regeneration projects shifted from local bureaucracy to city center business leaders in the course of conducting urban regeneration projects.

8.5.1 Abolishment of Urbanization Restriction Area

Kagawa prefecture abolished all the designated Urbanization Restriction Areas (URA) within the prefecture in 2004, and is the only prefecture to take this action so far. The prefecture government's intention was to reverse the trend in which population of central cities gradually decreases due to the designation of URA, but the population of suburban municipalities where URA is not designated increase rapidly. Sakaide city (population: 57,000 in 2005), adjunct to Takamatsu city and former member of the Kagawa Central Metropolitan City Planning Area along with Takamatsu city, had repeatedly requested the prefecture government to abolish URA in Sakaide city under the leadership of the mayor. The reason was to stop the population decline due to sprawling development in neighboring small municipalities, which were not included in the Kagawa Central Metropolitan CPA, and thus escaped URA designation (Fig. 8-7). In 2000,

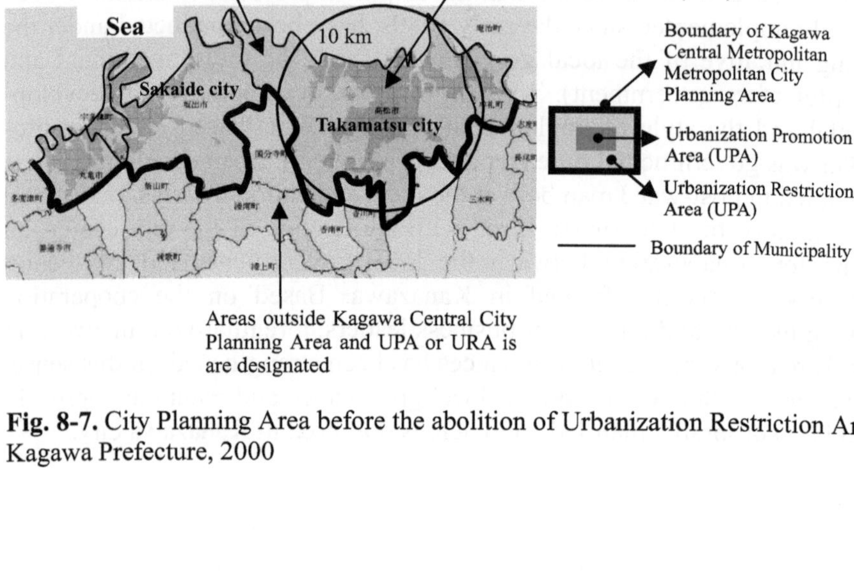

Fig. 8-7. City Planning Area before the abolition of Urbanization Restriction Area, Kagawa Prefecture, 2000

the City Planning Law was amended and the right to designate UCA and URA was devolved from the national government to the prefecture. After this amendment, Sakaide city again requested the abolishment of the URA designation, and Takamatsu city, whose population had started to decline, also joined the request. At the same time, the Kagawa Central Metropolitan CPA was dismantled and separate CPAs were created for Takamatsu city and Sakaide city, in response to a demand from Sakaide city's mayor. Thus, metropolitan considerations in city planning were definitely weakened as well.

8.5.2 Waterfront Urban Regeneration Project

The 13.1 km Seto Ohashi Bridge opened in 1988. The bridge connects Shikoku Island, where Takamatsu city is located, with Honshu Island, the main island of Japan. Takamatsu city had flourished as the gateway of Shikoku Island, since the Takamatsu port used to be a main port connecting Shikoku Island and Honshu Island by ferry. The Seto Ohashi Bridge does not link with Takamatsu city; thus, there was the fear that the role of Takamatsu city would be undermined. In conjunction with the opening of the Seto Ohashi Bridge and a subsequent close of ferry connection at the Takamatsu port, a waterfront development project (hotel, office building, shopping mall, convention hall, and new railway station) was undertaken as a symbol of Takamatsu city's new image. This project was implemented under the initiative of Kagawa prefecture and Takamatsu city with considerable subsidies from the national government (Fig. 8-8). In this sense, the waterfront development follows a bureaucratic project-led style typical of urban development in the post-war period. The project was grandiose and the demand did not match the scale of the project. Indeed, more than half of the developed plots are still vacant and office spaces in the symbol tower building are not well rented, which imposes a financial burden on the city and prefecture.

8.5.3 Shift in the Initiative of Urban Regeneration Project

Faced with unfavorable conditions for the waterfront development, recent initiatives for urban regeneration have shifted from local governments to the city-center retail business leaders. Urban redevelopment projects in the city-center shopping street (Marugamecho) are being conducted under the initiative of the Marugamecho Retailers' Association, including the conceptual stage through planning and implementation, as well as negotiations with the national government for subsidies. Local governments (both the city and prefecture) played only a supporting role. This is actually a very rare case in Japan. The members of the Marugamecho Retailers' Association are landowners in the Marugamecho central shopping district and many are retailers themselves, as is the usual case in Japan. But, the government sector also plays important roles in terms of the provision of subsidies. Indeed, one-third of project funds were subsidized by the government sector, half from the national government, and 25% each from the city and prefecture.

Takamatsu city is one of a handful of cities other than large metropolitan cities (Tokyo and Osaka) which are designated as an 'urban renaissance improvement area' and 'urban renaissance special district' under the Special Law on Urban Renaissance 2002. Through this designation, urban planning regulations were deregulated in the city center, making FAR bonuses possible (Fig. 8-8). This designation itself was also initially proposed by the Marugamecho Retailers' Association in direct consultation with the national government. The national government expected the case to be a model project for the promotion of regional cities through deregulation measures under the newly enacted Urban Renaissance Law.

As illustrated here, though local governments still play important roles in funding, it can be said that deregulation at the national level promoted a shift of the initiative for urban regeneration to the private sector (city-center business leaders). Given this situation, naturally, urban regeneration becomes market-oriented and project-based. Yet, it is not clear whether a few successful projects can become driving forces of regional development. It is more likely that it will create a few winners and many losers in

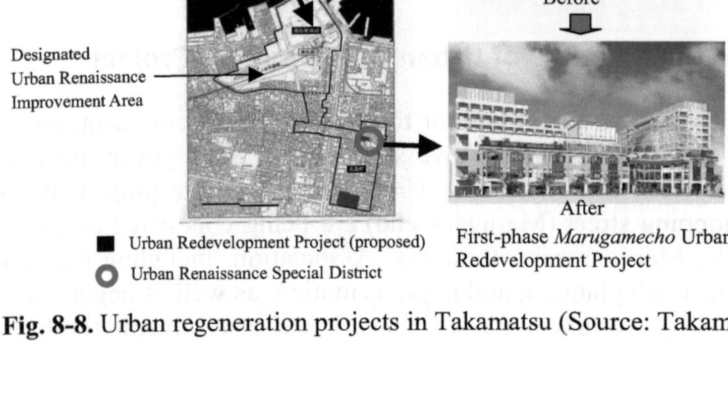

Fig. 8-8. Urban regeneration projects in Takamatsu (Source: Takamatsu City)

the region, given that there is no consensus on strategic vision to utilize successful projects for regional development as a whole.

8.6 The Case of Matsuyama City

Matsuyama city (population 515,000 in 2005), the capital of Ehime prefecture, employs a unique urban regeneration strategy that promotes various small-scale urban regeneration projects/events through the initiative of citizen groups. In this sense, Matsuyama city can be regarded as a typical example of the ad-hoc citizen initiative approach to urban regeneration.

8.6.1 *"Sakano Ueno Kumo" machizukuri*

Under the mayor's leadership, Matsuyma city is conducting a unique urban regeneration strategy named after a Japanese novel called *Sakano Ueno Kumo*. The theme of the novel is about young people born in Matsuyama who establish themselves. The mayor thought that this coincided with his idea of establishing a self-reliant city. From a realistic viewpoint, the novel was to be broadcasted as a drama by Japan's public TV broadcasting company (NHK), and thus it was expected to enhance Matusyama city's image. In that sense, this strategy has the nature of city marketing as well.

Against this background, an urban regeneration strategy called *Sakano Ueno Kumo machizukuri* (town planning) was started in 2004. The essence of the strategy is to promote ad hoc initiatives of retailers' community associations (shopping street-based retailers' associations) and other various citizen groups to revitalize the city by utilizing existing local resources (Fig. 8-9). In order to promote citizen group initiatives, the city has conducted a series of programs such as the citizen leadership training program and the financial support program for citizen groups to promote locally based participatory tourism. In fact, several locally based tourism projects have been generated through these programs. It is noted that, in the course of the conduct of these programs, a local intermediate NPO (non-profit organization) was also created by one of the citizen leaders and it is currently commissioned by the city to operate the citizen leadership training program. It can be said that the arena of collaboration between the city government and citizen groups is being generated.

As for the collaboration with retailers' community associations, a remarkable example is townscape improvement in the Ropeway shopping street in the city center. The Retailers' Community Association of the

Field Museum concept: Networking of various initiatives of revitalization activities by citizen groups for locally based tourism

Townscape improvement initiated by a retailer community organization

Fig. 8-9. Concept of *SakanoUeno Kumo Machizukuri* (Source: Matsuyama City)

Ropeway shopping street agreed to self-regulate building facade design in conjunction with the design improvement of the road by the city (Fig. 8-9). According to an interview with the leader of the Retailers' Community Association, consensus-building facade design became feasible only because the mayor showed a clear vision for the city development through the *Sakano Ueno Kumo machizukuri*. This case seems to illustrate that a clearvision and strategy is important to promote ad-hoc citizen initiatives.

Retailers' community associations are formed on a street-by-street basis in Japanese cities and are in competition with each other, thus often making it difficult to cooperate. In Matsuyma, a new enterprise called the Machizukuri Matsuyama company was created by city-center retailers as a new initiative to promote common management and marketing activities among separate retailers' community associations. This initiative is noted from the viewpoint of a new style of networking among retailers' community associations, though the outcome still remains to be seen.

Other bottom-up style attempts through the initiative of citizen groups have emerged. The *haiku* (Japanese traditional poetry) competition participated by a lot of high schools throughout the nation was initiated and organized by the Matsuyama JC (Junior Chamber, a young business leaders' association) and became one of the major events in Matsuyama city. The case of the preservation of the Koshin An House is also interesting in the sense that citizens and the city cooperated in various stages and deepened their collaboration. The Koshin An House used to be a hamlet of a local haiku poet, but was in a danger of demolition. This project was initiated in response to a proposal made by a college student during the city's citizen policy proposal program. After it was restored by the city in collaboration with a citizen group, the utilization of the preserved historic building as a haiku culture center was commissioned to a newly created NPO.

As seen in the above examples, the city development strategy of Matsuyama city is characterized by the concept of promoting the creation of new citizen groups and networking them through the support of citizen initiatives for bottom-up urban regeneration attempts. New citizen initiatives have been advocated by the current mayor as the basis for revitalizing the city. It should also be pointed out as political background that the mayor is an ex-member of Parliament and was rather free from the traditional bureaucratic structure of local governments.

8.7 Conclusion

This chapter proposes a conceptual framework for governance styles and urban regeneration planning approaches, which is used to analyze three Japanese regional cities and identify three typical approaches for urban regeneration. They indicate a shift from the post-war bureaucratic project-based approach for city development to collaboration among different sectors, reflecting the transition from government to governance.

Arena for Collaboration

New directions for urban regeneration will surely be generated from the collaboration among the government, private sector and civil sector. An arena must emerge for collaboration among the various stakeholders for creating a vision and strategy. In the case of Kanazawa city, an arena for collaboration is formed for the city center development among the city government and the city-center business leaders in the style of corporatism. However, it is noted that few citizen groups concerned with *machizukuri* (town planning) were created aside from the one created by the city-center business leaders. In Takamatsu city, the shift of initiative in urban regeneration from the city and the prefecture to business leaders of the city center is observed. In this case, business leaders cooperated directly with the national government, which is the largest investor in the urban redevelopment project through subsidies. Thus, a local arena of collaboration was not formed. Matsuyama's case is suggestive in that it has promoted the creation of citizen and business groups who are trying to create locally based urban regeneration with a certain degree of success. The emergence of a local intermediary NPO for networking citizen groups and an enterprise for common management and marketing of city-center retailers' community associations is believed to be an important step towards the evolution from an 'ad hoc' to 'networked' citizen initiative approach, though it is yet to see whether it succeeds in networking.

From Local Governance to City Region Governance

In all three cases examined here, clear mechanisms to make regional spatial development policy are not observed. On the contrary, deregulation in city planning led to the dismantling of the Kagawa Central Metropolitan City Planning Area and thus the coordination power of the prefecture at the metropolitan scale was significantly weakened in the case of Takamatsu. In Kanazawa and Matsuyama the CPAs have already been fragmented and do not cover actual metropolitan areas. The reason for this fragmentation is largely due to political objections from suburban municipalities refusing to be included in the metropolitan CPAs. The current urban planning system in Japan is quite weak when it comes to tackling regional issues. There is therefore a need for institutional settings at the regional scale through an amendment of the City Planning Law, in order to create an arena for governance at the regional scale.

References

Endo, K. (2006) "Case of Kanazawa city" (Presentation at the International Workshop on Sustainable Urban Regeneration and Local Governance, 17 January, 2006 at Tokyo, Japan)

9. Activities of Local Organizations for Downtown Revitalization in Regional Cities

Fumihiko Seta

9.1 Downtown Revitalization of Regional Cities in Japan from the Perspective of Local Organizations

One of the most serious problems at present in regional cities in Japan is the decay of downtowns, caused mainly by the acceleration of motorization and suburbanization. Though the phenomena are similar to the problems facing inner cities in most large cities in Western countries, the main concerns are different. While deterioration of public safety and segregation are among the main concerns for inner cities in Western metropolises, the primary problems confronting downtowns – and whole cities – in Japan's regional cities are depression of downtown shopping streets, degradation of the living environment, and the collapse of community identities.

In order to tackle downtown revitalization, not only public organizations like municipalities and prefectural governments, but also indigenous local organizations and newly established incorporated NPOs and voluntary organizations have been engaged in various activities and projects. Though there are many unique and significant activities implemented or enhanced by these organizations, the overall situation of downtown revitalization is difficult to grasp because the organizations and participants involved in downtown revitalization are complexly intertwined in each city. It is necessary to explain these at the outset for the assumed readers of this book, i.e. foreign scholars, researchers and businesspeople, most of whom are not familiar with the overall situation of downtowns and local activities in regional cities in Japan.

The chapter describes the situation of downtowns in Japan's regional cities, especially focusing on local organizations surrounding downtown shopping areas, and discusses their roles in connection with downtown revitalization and urban regeneration. Section 9.2 explains the overall conditions and constituents of downtowns as the basis for discussion in further sections, along with an overview of statistics about public attitudes toward downtown revitalization. Section 9.3 details a typical example of activities by local organizations related to downtown revitalization. Based on these descriptions, Section 9.4 concludes with necessary strategies for downtown revitalization and urban regeneration of regional cities, along with the ideal form of activities for local organizations.

9.2 Structure of Local Organizations and Activities in Downtowns

9.2.1 Public Attitudes regarding Downtown Revitalization in Japan

In most cities in Japan, people have enjoyed most public and retail services in central downtowns, supported by well-developed public transportation, geographically centralized urban areas and relatively strict regulations against large-scale retail stores in suburban areas. But the decline of downtowns has become increasingly obvious from the 1990s, when motorization and suburbanization accelerated and a law restricting the location of large-scale retail stores (Large-scale Retail Stores Law) was abolished.

While it is impossible to explain the details of the decay of downtowns in Japan in this section, public attitudes toward downtown revitalization are summarized here through the results of public opinion polls conducted by the central government in 2005, as this connects to local activities.

Fig. 9-1 shows public opinions regarding the ideal form of the city in the

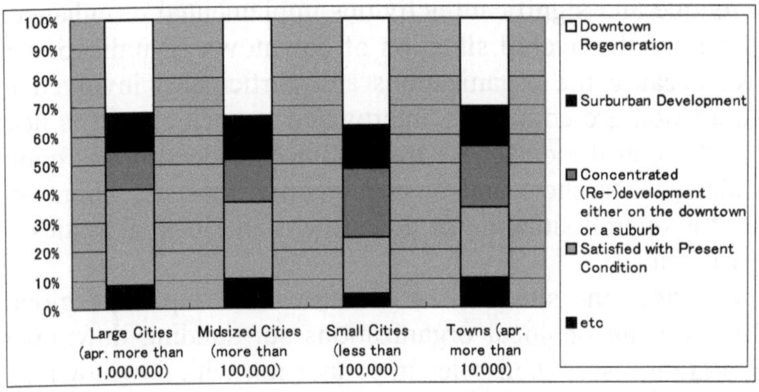

Fig. 9-1. Ideal form of the city in the future (Source: Public Opinion Polls 2005)

future. In all four ranges of cities, people tend to prefer downtown revitalization over suburban development. In particular, small cities (with population of approximately 10,000 to 100,000) have the smallest ratio of people who are satisfied with present conditions of their cities. They tend to desire either downtown revitalization or concentrated (re-)development in their downtown or a suburb.

Fig. 9-2 shows desirable roles or wishes for downtowns, grouped according to four ranges of cities. There are some differences regarding desirable roles and wishes among the city ranges, but they all point to the supply of comprehensive services, supply of basic necessities, the role as a community, supply of lifestyle support services for the elderly, and reinforcement of public transportation. The results imply that downtowns are perceived not only as places that supply daily commodities, but also places to fulfill various roles and functions like public services and being a neighborhood.

There are many similar surveys on downtown revitalization conducted

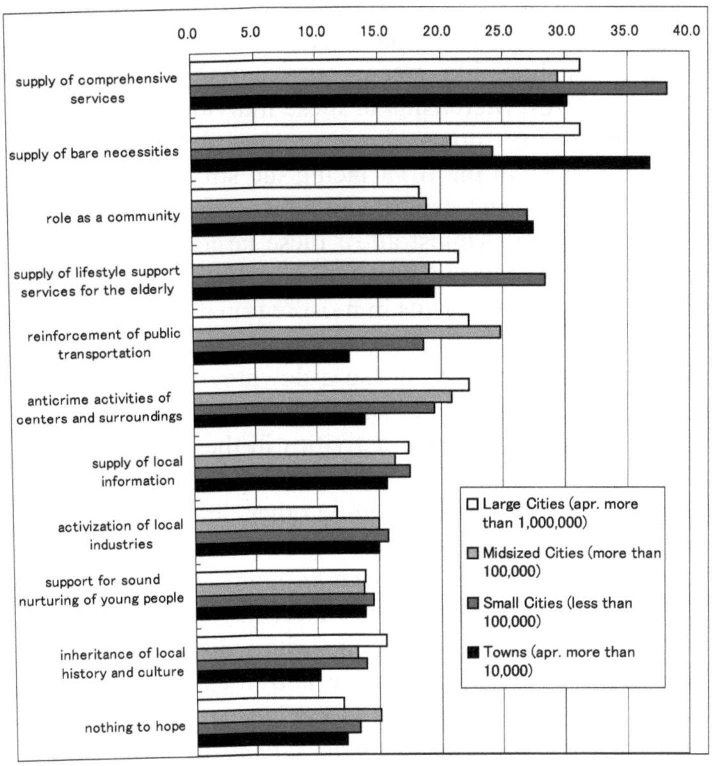

Fig. 9-2. Desirable roles or wishes for downtowns (Source: Public Opinion Polls 2005)

by public and private bodies, most of which commonly affirm the importance of downtowns for citizens in regional cities, especially for the elderly. Thus, downtown revitalization can be regarded as an essential issue, not only for municipalities or downtown merchants, but also for citizens.

9.2.2 Outlook of Local Organizations in Downtowns in Japan

Here the general situation of local organizations in typical downtowns in regional cities in Japan is described in order to clarify the complexity of local activities and the close connections to average residents.

Regional cities in Japan were in the past shaped by their vigorous downtowns in the city center, consisting of shopping streets and some public facilities like the city hall, post offices, community centers, etc. supported by public transportation terminal like trains and buses. Shopping streets are in the center (downtown) of the city, surrounded by residential areas, and form the urban structure of the city as a whole. People come to downtowns on foot or by bicycle, bus, tram or train and buy daily commodities and enjoy various services. A downtown shopping area was formerly composed mostly by small individual shops like greengroceries, fish shops, butchers, drugstores and clothing stores. Merchants of these individual shops generally live in the shop-houses on upper floors or toward the back of the ground floor, and do business in the front facing the street. Later, a business category of supermarkets and department stores emerged and joined the composition of downtowns. Almost all of these large retailers located in or near a downtown --- and not the suburbs --- because of legal regulations. Thus, the popularity of downtowns in most regional cities was maintained until 1990s, even if some declined.

There are various entities which compose a downtown shopping area. Shopping streets in a downtown form local merchants' associations (*shotenkai*) for management and cooperative activities in the street. Some are called promotion associations (*shinko-kumiai*) or cooperative associations (*kyodo-kumiai*) based on respective laws, while others are voluntary associations. The chamber of commerce and industry (CCI, *shoko-kaigisho*) or the societies of commerce and industry (*shoko-kai*), one of which is established in almost every municipality, play an important role in local activities for revitalization a shopping street, or the whole downtown in some cases. In addition, some downtowns feature activities of citizens and commuters who are not necessarily related to the businesses on the shopping streets. Neighborhood associations (*chokai*, *chonaikai* or *jichikai*) are typical indigenous local organizations. Recently the number of incorporated nonprofit organizations (incorporated NPOs) has been steadily in-

Fig. 9-3. Downtown shopping area in Japan (Kumamoto City)

creasing based on the NPO Promotion Law (1998), and many of these are engaged in offering various public services. Also, several other kinds of organizations or meetings formed for a particular purpose often affect the activities of downtowns, such as redevelopment cooperatives and Town Management Organizations (TMOs) designated for downtown revitalization.

In many cases, merchants in a shopping street and some downtown residents become members or even executives of multiple organizations. Some chairpersons of local merchants' associations double as executives of neighborhood associations, while a staff person on the CCI who commutes from another city becomes one of the leaders of a voluntary organization in the downtown.

The following sub-sections explain each local downtown organization.

9.2.3 Local Merchants' Association (*Shotenkai, Shinko-Kumiai, Kyodo-Kumiai*)

A local merchants' association consists of most of the retailers in a shopping street. Local merchants' association (*shotenkai*) is the collective term, and some are called (shopping street) promotion associations ((*shotengai-*)*shinko-kumiai*) or cooperative associations (*kyodo-kumiai*), based on the Shopping District Promotion Association Law (1962) and Law on the Cooperative Association of Small and Medium Enterprises (1949), and they can receive some support and subsidies from public bodies for the promotion of their street. The other local merchants' associations are voluntary (*shotenkai*, in the narrow sense), but in some cities they are registered through ordinances or outlines and supported by municipalities.

Since downtown shopping areas in Japan are usually strung out along the street, more than one local merchants' associations are established on each street, even in a small downtown. These associations often form one or several federations in a part of or the whole downtown, and have cooperative relationships.

Local merchants' associations are engaged not only in collaborative business activities like group purchasing, but also in maintaining shopping street equipment and facilities such as benches, plants, billboards, streetlights and roofs in a Japanese-style arcade, and in group activities like downtown events and distributing brochures. Costs for these activities are covered by membership fees and some subsidies from public bodies. Though it is not compulsory for merchants or landowners to become members of an association, unlike a BID (Business Improvement District, often seen in the US), most merchants participate because it seems difficult to manage without collaborating with these associations in the shopping street.

According to a survey in 2003 conducted by the Federation of Shopping Center Promotion Associations, there are some 13,000 local merchants' associations in Japan, and about one-fourth are promotion associations, one-tenth are cooperative associations, and the rest (around three-fifth) are voluntary associations. Each association consists of around 50 members,

Fig. 9-4. Typical arcaded shopping mall in Japan (Kochi city)

and on average more than 80% of merchants in a shopping street are members or associate members.

9.2.4 Chamber or Society of Commerce and Industry, and Junior Union (*Shoko-Kaigisho, Shokokai* and *Seinen-Kaigisho*)

The chamber of commerce and industry (CCI, or *shoko-kaigisho*) and the societies of commerce and industry (*shokokai*) are designated respectively by the Chamber of Commerce and Industry Law (1953) and Commerce and Industry Association Law (1960), and one of them is established in every city. Generally, chambers are set up in large cities and societies in small towns, but their activities are not so different. The chamber or society provides business, information and welfare support which cannot be easily obtained by individual small merchants. Since a chamber or a society is specified in each city, their activities and services are usually highly localized to their city or town and closely connected to other organizations in the downtown in diverse ways. Regarding relationships with individual merchants in a downtown, the staff of the chamber or society usually offer consultation services, support various applications and declarations like income tax returns, and sometimes help businesses obtain small loans. In addition, sections of the chamber or society such as youth or female groups voluntarily engage in various kinds of local activities to promote the industries in the city.

As of April 2006, there are 517 CCIs in Japan with a total membership of 1,450,000, and 2,342 societies of commerce and industry with a total membership of 1,200,000. The sum of CCIs and societies exceeds the number of municipalities, since they both have organizations at the prefectural and national (central) level. The central chamber and society carry a lot of weight for national economic and industrial policies in Japan.

From the view of the constituent members, CCIs and societies are established to support not only downtown merchants but also any other kind and scale of industry and commerce in the city; thus, they do not necessarily have to support downtown revitalization. For some CCIs membership includes strong suburban GMSs (General Merchandise Stores), whose interests oppose downtown revitalization. On the other hand, CCIs and societies are basically just for promotion of industry and commerce, without direct relationships with citizens in the city. However, CCIs and societies often play a principal role in downtown revitalization in many ways, such as becoming the designated TMO (Town Management Organization) as explained later.

Junior Chambers (JCs, or *seinen-kaigisho*) are also organizations related to local economy and downtown revitalization, although their role is different from CCIs or societies. JC members are merchants or managers from any industry in the city who are 40 years old or younger and are engaged in local political, economic and social activities, some of which could be related to downtown revitalization.

9.2.5 Incorporated NPO and Voluntary Organization

The term nonprofit organization (NPO) in a broad sense refers to any organization that does not earn profits and performs various activities often by voluntary citizens motivated to contribute to society or their community. In Japan, for a long time there was no suitable corporate status for these voluntary organizations. It was almost impossible for these organizations to gain formal status because of cumbersome procedures and strict standards. Thus, most management tasks were conducted under the names of individual staff members.

In order to solve the problem and respond to the expansion of voluntary activities in Japan, the Nonprofit Activities Promotion Law (NPO Law) was enacted in March 1998, and the new legal status of 'incorporated NPO' was created for voluntary organizations that contribute to society or their locality and fulfill certain requirements. The application process is much easier than other types of legal bodies, and if certified, the organization can own property and sign contracts under the organization's name.

The 17 categories of activity specified in the NPO Law include many related to downtown revitalization, such as improvement of cities, economic vitalization, social education, welfare activities, environmental conservation and community safety. As of June 2007, more than 30,000 organizations have been designated as incorporated NPOs.

In addition to these incorporated NPOs, various voluntary organizations are directly and indirectly concerned with promoting shopping areas and downtown revitalization. Their activities include holding various local events, extension of community (small) businesses, operating child-care and nursing-care facilities, and networking among these local activities and organizations.

9.2.6 Neighborhood Association (*Chokai, Chonaikai* or *Jichikai*)

Neighborhood associations (*chokai*, *chonaikai* or *jichikai*) are voluntary organizations maintained by residents in each neighborhood, separated by postal boundary (*chome*), school district or street. They are quite common as the most basic local organization in every residential area in Japan. Neighborhood associations collect small membership fees from residents, and supply various services like distributing local information by circular, holding local festivals and events, supporting members' ceremonies (often funerals), cleaning their district and streets, contributing to local accident- and crime-prevention, and maintaining their neighborhood public facilities. Most activities are performed by residents themselves. Though the associations are generally private organizations and are not directly related to the municipalities, many handle aspects of public services as described above, and in some cities neighborhood associations are virtually regarded as representatives of residents.

However, the situation regarding neighborhood associations has been gradually changing in recent years. There are more residents, especially in metropolitan areas, who don't participate in the association and don't pay the membership fees. Meanwhile, some neighborhood associations, especially in areas impacted by residential redevelopment, limit the rights of new members to protect their vested interests. Thus, the situation of neighborhood associations has become varied and unstable.

In terms of the relationship with downtown revitalization, the activities of neighborhood associations are also often connected to specific events such as local festivals, like those of incorporated NPOs and voluntary organizations. Since neighborhood associations are usually made up of residents in the district or street, most merchants of a downtown shopping street, who are now or were in the past residents of the street, are also members of the association. Therefore, neighborhood associations often play an active role in downtown revitalization.

9.2.7 Others (Organizations or Groups for a Specific Purpose)

In addition to the organizations described above, some local organizations convene for a specific purpose or project. Since promotion of shopping streets and downtown revitalization is regarded as an important task in a city, still more special organizations or groups related to a downtown tend to be established. In most cases, their leaders are also leading members of above-mentioned local organizations.

Redevelopment Cooperative

Downtowns in regional cities in Japan are very dense with many wooden buildings, and are quite vulnerable to earthquakes and fires. Thus, many blocks in downtowns are designated as target areas for land readjustment projects or urban redevelopment projects to improve resistance to such disasters. Moreover, when a large-scale store like a supermarket is closed and few tenants can be found, mixed-use redevelopment tends to be planned collectively by the municipality, landowners and related local organizations in the downtown. A redevelopment cooperative is organized by these participants to make a suitable redevelopment plan and invite tenants after the completion of construction. Since the result of a downtown redevelopment project strongly affects the success or failure of the shopping area, the cooperative's decision-making is quite important for downtown revitalization.

Town Management Corporation "Machizukuri-Gaisha"

A town management corporation (*machizukuri-gaisha*) in a broad sense means a corporation engaged in any kind of business in a town, but in general it refers to a corporation which is funded by the municipality, other public bodies and many merchants or residents, and is involved in various activities and their management for a certain town or district. For those located downtown, the main revenue is gained through parking fees, and finances are bolstered by municipal support and subsidies, so that most of them are not financially independent. Some downtown management corporations are designated as Town Management Organizations (TMOs), explained next.

Town Management Organization (TMO)

Town Management Organizations, set up formally by the Central City Invigoration Law (1998), work toward downtown revitalization through comprehensive management and operation of various activities mainly in downtown shopping streets and areas. In each municipality, either the chamber or society of commerce and industry, or public corporation like a downtown management corporation, is designated as a TMO based on the TMO Vision, which states the desirable future of the downtown and its roadmap. Once a TMO is selected in a downtown, it becomes the main recipient of public support and subsidies for downtown revitalization, and conducts and coordinates various activities and projects in the downtown, performed by local organizations or by itself.

The TMO system was established based on the common recognition of the lack of integration among activities and functions in most of the downtowns in Japan's regional cities. Many downtown activities conducted by individual merchants, local merchants' associations, the chamber or society of commerce and industry and other local organizations are not sufficiently integrated or well-managed. TMOs aim at downtown revitalization by integrating these activities.

However, downtown revitalization is not realized in most regional cities at present, regardless of the establishment and operation of TMOs. Specialists have pointed out several intrinsic weaknesses, including lack of human resources to manage and integrate various individual activities, shortage of independent revenue sources, and insufficient authority for coordination.

The Central City Invigoration Law was amended in 2006 and the position of TMOs as recipients of public support and subsidies has diminished. However, TMOs remain important for downtown revitalization in some cities.

Central City Invigoration Council

Due to the amendment of the Central City Invigoration Law in 2006, the formal recipient of public support for downtown revitalization was changed from TMOs to Central City Invigoration Councils, which consist of various local organizations related to a downtown. The councils are to conduct and coordinate various activities for revitalization. The amendment requires two types of members within the council. One is for urban development in the downtown, e.g. a public corporation for urban development, and the other is for promotion of downtown commercial business, e.g. the chamber or society of commerce and industry or a town management corporation. But there are many other optional members related to downtown revitalization like the municipality, individual merchants, local merchants' associations, financial institutions (banks and local credit unions), local incorporated NPOs, other voluntary organizations, (semi-)public corporations, local media firms (local newspapers, community FM, etc.), and so on. These members promote comprehensive downtown revitalization through the council.

Other meetings and workshops

Voluntary meetings and workshops are often organized for events, preparation for redevelopment, invitation of new tenants, and reutilization of vacant spaces and stores. Most consist of municipal officials and leading members of the local organizations described above.

9.2.8 Roles of Local Organizations for Downtowns Revitalization

While individual local organizations perform their activities based on their original purposes, capacities and motivations, achieving downtown revitalization in Japan requires the assignment of roles to local organizations. TMOs were established in many downtowns as bodies to coordinate these activities. However, most of them are still not successful enough to achieve revitalization of downtowns and shopping streets in regional cities.

In the meantime, there are several downtowns regarded as successful cases of revitalization. In most cases, revitalization proceeds through consensus among various local organizations and coordination of various activities, even though there are some famous and symbolic key persons either in the public or private sector. In that sense, vigorous activities by local organizations is a necessary condition, but not a sufficient condition. Rather, the appropriate coordination body seems to be essential.

Another important point regarding revitalization through local organizations is the structural change of each downtown. As mentioned before, most merchants in shopping streets previously lived there and acted not only as merchants but also as residents. Hence, the interests of different organizations (such as, local merchants' associations, neighborhood associations and other voluntary organizations) were largely the same or similar. However, in keeping with changes in the Japanese economy, individual merchants turned a large profit and tended to construct or purchase their new residences in the suburbs. As the tendency became stronger, downtown shopping streets started to be regarded merely as a place for commerce, and that perception made it more difficult to manage local activities comprehensively. The tendency has become stronger since the decay of downtowns in recent years. Even though other businesses want to rent vacant spaces or shops, owners tend to refuse offers just because they don't feel the necessity, and leave them vacant or closed. On the other hand, when a landowner of a plot along a downtown street sells vacant space to a residential developer, a large condominium tends to be constructed in the center of the shopping street. Many of the new residents are related neither to commerce nor to downtown activities for revitalization, so they are not concerned with the decay of downtown shopping streets. All of these phenomena have accelerated the decay of downtowns.

Given the recent trends described above, it is all the more important to promote individual activities undertaken by various local downtown organizations, and also to form a system in which these activities can be effectively managed for revitalization from an integrated standpoint.

9.3 Roles of Local Organizations in a Downtown: The case of Toyama City

As introduced in Chapter 10, Toyama City is a regional city along the Sea of Japan, separated from major metropolitan areas. Suburbanization has accelerated along with motorization and downtowns have drastically declined in recent years due to the expansion of suburban GMSs. In response, various local entities have tried to revitalize downtown shopping streets by reforming the urban structure through various measures related to 'compact city policies' initiated by the Toyama municipal government.

Since municipal policies are explained in the other chapter, this section picks up some typical activities for downtown revitalization by local organizations in Toyama and clarifies the potential and future tasks of these activities, most of which are common to downtowns in regional cities.

9.3.1 Machizukuri Toyama Inc.

In July 2000, Machizukuri Toyama Inc. was established as the town management corporation for downtown Toyama. The Master Plan under the former Central City Invigoration Law (1998) was enacted in 1999, and the corporation was designated as the Town Management Organization (TMO) in order to coordinate various activities in downtown Toyama.

Fig. 9-5. The amenity facility Forza Sogawa operated by Machizukuri Toyama Inc.

Undertakings include existing local organization activities like the 'challenge shop' project described below, as well as newly started projects for downtown revitalization. New projects include operation of an incubation office, amenity and neighborhood facilities, maintenance of the tram and other public transport, holding various events like festivals and bazaars, and distributing local information.

There are 38 shareholders as of 2007 and half of the capital is funded by Toyama City, 16.7% by the Toyama Chamber of Commerce and Industry (the Toyama CCI), 20.0% by several local commercial or industrial associations and individual enterprises, and 13.3% by other sources (large-scale merchants, banks, etc.).
The corporation is operated through board meetings, made up of representatives from local merchants' associations, private companies, the Toyama CCI, and Toyama City. Under the direction of the meetings, the project planning and promotion department and administrative department conduct various operations as described above. In addition, multiple platforms for local residents are set up to tap a wide range of opinions about the operation from all levels.

Revenues for 2007 (budget base) are composed by each one third of parking revenue, of subsidies mainly by Toyama City, and of others; thus, it is not self-supporting.

Exceptional activities can be seen in some specific projects like the operation of a tram and the 'challenge shop' project, while collaboration among local organizations for decision-makings and operations, strong public-sector financial support, and dependence on parking revenue are typical of most town management corporations in regional cities.

9.3.2 "Challenge Shop" Business

"Challenge shops" mean vacant stores or portions of stores rented to new small business entrepreneurs at a low rate, or at no cost, usually for a limited term. In order to revitalize commercial downtown business, local merchants' associations or municipalities in some regional cities open challenge shops and invite new commercial tenants. Most of challenge shops were formerly shops or shopping centers which were closed due to the decline of downtowns, so local downtown merchants' associations or municipalities regard challenge shops as an effective way both for reclamation of vacant shops and for cultivation of human resources in downtowns. In most cases, new tenants are screened based on the feasibility of their business, the existing tenant mix, and their social contribution to the downtown or the whole city. In addition the rent subsidy, the local merchants' association and other organizations offer selected challenge shop tenants various kinds of support like business consultation, collaboration with other merchants, and networking. These new tenants are expected to 'graduate' from challenge shops after the limited term to an independent business, preferably in or near the same downtown.

Toyama is famous for a successful challenge shop project called Freak Pocket, and many cities in Japan try to create similar challenge shops in their downtowns.

Freak Pocket is said to be the brainchild of sisters who operated a variety store in a shopping street in downtown Toyama. They hit on the idea when they traveled to Hong Kong and observed small unique shops densely packed in small spaces. After returning, they reported their observations and introduced their idea to their local merchants' association, and it was realized as the Central Street (Chuo-Dori) Mini-Challenge Shop Plan supported by Toyama City. Then several small lots of approximately 7 square meters, separating spaces of a two-story building, were rented for a limit of one year to promising new tenants, selected through applications and interviews about their funding ability and commercial potential. After these tenants came to Freak Pocket, workshops and on-the-job training were held by certified commercial specialists like SME (Small and Medium Enterprise) consultants, tax accountants and commercial space designers, dispatched through Toyama City, and young merchants from the local merchants' association. Communication and information exchange among these new tenants were also facilitated.

Freak Pocket was at first equally funded by the local merchants' association and Toyama City, but the management was succeeded by Machizukuri Toyama Inc. after it was designated as the TMO, and was continued until 2005, when it was followed by other challenge shops in thedowntown.

As a result, around 70 tenants 'graduated' from Freak Pocket and 50 of them opened their own businesses. Many opened their own shops around the downtown and have made some contributions to downtown revitalization. Among these graduated shops, there are many clothing and variety stores for younger people and these shops attract young customers in Toyama, most who usually have lost interest in the downtowns of regional cities. Toyama City supports these shops even after they graduated from challenge shops through preferential loans.

The success of Freak Pocket is highly instructive for local organizations aiming at downtown shopping district revitalization. The project was initiated by a local merchant and realized through a local merchants' association, and strong support was continued by the municipality and further promotion was done through the TMO scheme set up by the central

Fig. 9-6. Shop that graduated from Freak Pocket

Fig. 9-7. East half of Chuo-Dori downtown shopping street (Sunday afternoon)

government. It is a typical case of successful collaboration for downtown revitalization, centered within local organizations and supported by the public sector.

However, the street (Chuo-Dori) that initiated Freak Pocket is still declining and has nearly decayed, like many downtown shopping streets in other regional cities. Taking into account that it is still difficult to revitalize downtowns even by initiating and operating such a successful project, it can be said that downtown revitalization can be realized only through multiple combinations of successful collaborative projects and activities held by various organizations.

9.3.3 Various Activities by Local Merchants' Associations

Local merchants' associations play a main role in promoting a downtown shopping area. In Toyama, Sogawa, Chuo and Nishicho streets are three main downtown shopping streets where several local merchants' associations, including promotion and cooperative associations, are engaged in various collaborative activities for revitalization. Unique activities include establishing a free parking day in the downtown area for suburban customers by tying up with surrounding parking lot owners, and publishing a book called *Sogawa Story* (*Sogawa Monogatari*) that reminds middle-aged and elderly customers of the crowded downtowns in the past. Additional projects are planned along with the relocation and renewal of Daiwa, a local department store.

9.3.4 Further Tasks for Downtown Revitalization

In spite of the variety of the activities by local organizations described above, it seems to be almost impossible to achieve revitalization of the entire downtown of Toyama. The downtown's former festivity will never fully recover, even if the downtown Master Plan goals under the former Central City Invigoration Law were wholly realized by the target year (after five years). Indeed, it is not difficult to achieve numeric targets (number of pedestrians, etc.) mainly through redevelopment and renewal of a local department store, but the standard of these targets is much lower than the past situation.

One of the largest obstacles to revitalizing downtown shopping streets through local merchants' associations is realizing an appropriate tenant mix. Their competitors, mainly suburban GMSs, offer an attractive composition of tenants by ejecting useless ones and inviting new appropriate ones in their gigantic suburban stores. Their top-down decision-making and clear contractual relationships with tenants enable suitable tenant mix. On the other hand, relationships among individual merchants in downtowns tend to be parallel, meaning they are at the same level. Even if the local merchants' association plans a preferable tenant mix for their shopping street, the realization depends on the decisions of merchants who own or rent their shops.

This parallel structure is problematic also when the shopping street tries to scale down and focus their resources on a certain area or sector. Even now, only the central part of downtown Toyama seems to be regenerable, while other peripheral parts are already or about to decay. Even under the circumstances, it is difficult to reconcentrate preferable shops spread in the downtown and focus their resources for improved competition. In the downtown, individual merchants or shopping streets have relationships that are both collaborative and competitive. Coordination of interests within complicated relationships cannot be easily achieved given the diminishing equilibrium.

At any rate, strong leadership along with consensus among all entities in the downtown are necessary for planning and realizing effective revitalization strategies. Total collective and collaborative effort in the downtown is put to the test.

9.4 Conclusion

As summarized in Section 9.2, local organizations in downtowns in Japan are quite varied, i.e. local merchants' associations, neighborhood associations, the chamber or society of commerce and industry, up-and-coming incorporated NPOs, various voluntary organizations, as well as the municipality. These organizations are now collectively struggling to resist suburbanization, revitalize downtowns and realize a compact city in many regional cities in Japan.

Section 9.3 introduced several examples of downtown revitalization activities in Toyama. Like Toyama, the town management corporation and challenge shops are commonly seen in many downtowns in Japan, but in many cities these efforts seem to be sluggish due to differences in external and internal conditions.

Downtown revitalization cannot be accomplished by independent activities, even if they are quite unique and attractive. Downtown revitalization and urban regeneration of regional cities in Japan will become a reality only if wide-ranging and vigorous activities by local organizations are comprehensively coordinated through strong leadership and sufficient consensus.

References

Central Federation of Societies of Commerce and Industry, http://www.shokokai.or.jp/ (Accessed in 2007)

The Japan Chamber of Commerce and Industry, http://www.jcci.or.jp/ (Accessed in 2007)

Machizukuri Toyama Inc., http://www.tmo-toyama.com/ (Accessed in 2007)

Ministry of Economy, Trade and Industry, http://www.meti.go.jp/ (Accessed in 2007)

Public Opinion Polls 2005, http://www8.cao.go.jp/survey/h17/h17-kouri/index.html (Accessed in 2007)

Junior Chamber International Japan, http://www.jaycee.or.jp/ (Accessed in 2007)

10. Sustainable Regeneration of a Car-dependent City: The Case of Toyama toward a Compact City

Kiyoshi Takami and Kiichiro Hatoyama

10.1 Introduction

Toyama Prefecture is one of three prefectures that comprise the Hokuriku Region at the Sea of Japan side of central Japan. Toyama City is its prefectural capital, and the second-largest city in Hokuriku after Kanazawa City (Fig. 10-1). About 420,000 people live in the city after the merger of seven municipalities in 2005, and the population of its metropolitan area[1] is about 540,000.

From a geographical viewpoint, Toyama prefecture was historically vulnerable to floods and mudflow attacks, since the prefecture has many torrential rivers, as well as the huge Tateyama caldera (natural volcanic formation). Due to these geographical conditions, Toyama prefecture has acknowledged the importance of flood and erosion control. For example,

Fig. 10-1. Location of Toyama City

the construction of a cut-off channel (1901) to prevent the center of Toyama City from floods by snaking the torrential Jinzu River and the erosion control work (1906) against mudflow from the Tateyama caldera were famous engineering enterprises in Japan.

From an industrial viewpoint, the prefecture has grown with industries such as refining and processing of aluminium, machine industry and medicine manufacture, along with agriculture and fishery. It often has a reputation as the most livable prefecture in the country, which may derive from the highest homeownership rate in Japan (79.1%, according to the National Census 2005), the highest average house size, a rich natural environment, a relatively high prefectural income per capita, and so on.

Toyama City, however, has experienced suburbanization and hollowing of the city center, as with many of Japan's local cities, and is lagging behind Kanazawa in interurban competition. In the midst of recent trends of aging, declining birthrate and downturn in population, the city has started to shift its policy direction toward creating a compact city. The city is now attracting attention as it successfully opened the first Light Rail Transit (LRT) line in Japan in 2006 and gained the first authorization of its Central City Invigoration Plan from the central government in 2007.

This paper outlines the challenges which Toyama City has been facing and examines its ongoing compact city policy, focusing in particular on the successful introduction of LRT, and discusses issues on the policy and its framework.

10.2 Challenges Confronting Toyama City

Toyama City is now seeing several significant demographic, development and transportation-related challenges, summarized as follows.

- Aging and downturn in population: The city's population has been generally flat since the late 1990s, except for an increase due to the municipal merger. It is predicted to start declining after peaking in 2005 and decrease by about 10% over the next 20 years. According to the National Institute of Population and Social Security Research (2003), the ratio of those 65 and older will increase from 19.7% in 2000 to 27.7% in 2015 and 30.5% in 2030. These levels are almost comparable with the national average.
- Suburbanization and decrease in population density: The trend for population distribution has been a decrease in the central city and many outer-suburban areas outside the former Toyama City, and an increase between those areas. Population density in the city's Densely

Inhabited District is 40.3 persons/hectare in 2005, the lowest among all prefectural capitals in Japan. Such low-density suburbanization is seen as being problematic, as it will increase the administrative cost per capita by an estimated 12% in 20 years, including costs for snow clearance, street cleanup and maintenance of parks and the sewer system.

- Decline of the city center: Population in the city center dropped by half over the past 40 years. During the past decade, it decreased by more than 10%, although it may have hit the bottom with a slight increase in 2006 from the previous year, while the number, as well as the ratio, of those 65 and older has recorded a steady increase. It is said that elderly people have been moving to new apartments in the central area partly to be freed from the burden of snow clearance work. Its status as a central shopping district deteriorated with a 40% decrease in retail turnover from 1994 to 2004, compared with a 5% decrease for the whole of former Toyama City. The number of pedestrians on Sundays also dropped to one-third from 1995 to 2006 and the trend is continuing. These may be partly because of economic depression, but some shoppers have surely run off to suburban stores and the Kanazawa city center, which led to closure of stores, including Seibu Department Store in March 2006. The outflow of shoppers may escalate with more severe interurban competition after the opening of the Shinkansen bullet train (Hokuriku Shinkansen) for Tokyo and Kanazawa in 2014.
- High reliance on automobiles: One characteristic of Toyama Prefecture is the second-highest car ownership rate in Japan (1.73 vehicles/household in 2005). Car use has increased as car ownership rate has grown, combined with low-density suburbanization. According to a Person Trip Survey in 1999, 72% of all trips and 84% of commuting trips were made by automobiles, which is the highest among all local hub cities in Japan. Public transportation ridership has seen a dramatic decrease, and in particular bus ridership dropped to one-third from 1989 to 2004. Meanwhile, those who cannot use cars freely, defined by the Toyama City government as those who have neither a driving license nor one's own car, account for about 30% of the population 15 and older, 76% of which are female and 71% of which are 60 and older.

In the above context, the Toyama City government recognizes the following problems which will deepen as the population ages and decreases:

- inconvenience for those who cannot use cars freely,

- relatively high administrative costs to maintain infrastructures and provide city services, and
- deterioration of the city's vitality and attractiveness derived from the hollowing of the city center.

It may be worth pointing out that the city government thinks environmental sustainability issues are less problematic at the city level, and puts little emphasis on them, in particular the necessity to reduce car use from the environmental viewpoint.

10.3 Compact City Policy of Toyama City

In order to deal with the problems identified above, the Toyama City government is steering its urban policy toward compact city under the leadership of mayor Masashi Mori. The basic principle is described as 'urban structure consisting of sticks and dumplings,' where 'sticks' mean trunk public transportation lines and 'dumplings' mean walkable communities along them (Fig. 10-2). It is not a monocentric compact city structure, but rather a transit-oriented-development (aka TOD) type of structure. One factor is that the city has many rail and tram lines compared with other medium-sized cities in Japan. Another seems to be the difficulty of adopting a monocentric shape to the new city which was just merged with neighboring municipalities to become six times bigger in area.

The city government aims to realize this policy through the following three categories of measures, as shown in Table 10-1.

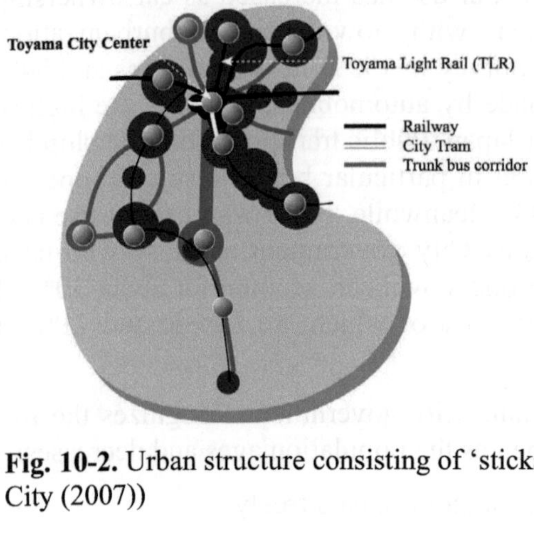

Fig. 10-2. Urban structure consisting of 'sticks' and 'dumplings' (Source: Toyama City (2007))

Table 10-1. Main measures for realizing compact city

Vitalization of public transportation	**Networking LRT and tram**	• Toyama Light Rail (TLR): the first LRT line in Japan • Experimental increase of frequency for the JR Takayama Line • Loop-line service for the city tram* • Run-through service on TLR, city tram and a suburban rail line
	Improving bus	• Improvement of trunk bus vehicles and bus stops • Community bus service to assure mobility
	Support for the elderly	• Fare discount for intracity trips to/from city center* • Support for surrender of driving licenses
Regulating and guiding developments	**Regulating developments**	• Regulation on trip-attracting developments in light-industrial zones and loose regulation areas • Regulation on sporadic residential developments in suburban areas
	Guiding population distribution	• Financial support for constructing, acquiring and renting houses in the city center and 'dumplings' along public transportation corridors** • Support for elderly households to move into the city center**
Invigoration of city center	**Making public transportation more convenient**	Target: Increase ridership of city tram by 30% in five years • Elevation of railway lines around Toyama Station • Land readjustment in the area around Toyama Station • Maidohaya community bus service to support intracentral area trips • (See also single-asterisked measures above)
	Creating lively city core	Target: Increase pedestrian flow in the city center by 30% in five years • Renovation of Toyama Castle Park • Development and operation of several key facilities • Mixed-use redevelopment projects including retail and residential • Refurbishment of exteriors under a unified concept • Special deregulation zones for attracting major retail developments • City Center Thanksgiving Day: free parking and celebratory events
	Promoting residence in the central area	Target: Increase population of the central area by 10% in five years • Several redevelopment/development projects including residence • (See also double-asterisked measures above)

10.3.1 Vitalization of Public Transportation

All railway and tram lines in the city and selected bus lines which run frequently (more than about 60 services/day), between the city center and district centers, and between the city center and important facilities (including university, hospitals and airport), are designated as public transportation corridors in the city's master plan.

As for rail and tram, the city government has so far opened Toyama Light Rail (TLR), the first LRT in Japan, and has conducted a social experiment of increasing frequency of the JR Takayama Line. Additionally, the city plans to network LRT and tram lines, providing a loop-line service of city tram and run-through services on TLR, city tram and a suburban rail line.

TLR was, as described in the next section, converted from the former JR Toyamako Line and opened with new vehicles and improved service in April 2006. Ridership has doubled on weekdays, tripled or quadrupled on holidays, and remains at a high level. A questionnaire survey in October 2006 shows that more than 80% of inhabitants in any district of the city, as well as in the district along TLR, highly appreciate TLR. The city government apprehends this is partly because the mayor explained heartily to citizens the necessity to shift toward a compact city and the importance of TLR as a leading project for the policy's realization.

They have, however, had a hard time in the social experiment at JR Takayama Line which started in October 2006. Under the experiment, train frequency is increased by half, which is nearly equivalent to 30-minute intervals in peak hours and 60-minute intervals in the daytime, along with feeder bus and shared-taxi services at several stations at the expense of Toyama City. But ridership during the first five months showed just a slight increase (+1.3%) from the previous year. In the second stage of the experiment starting from March 2008, the city government plans to double service frequency in the daytime, while shortening the experiment section to half, and to place a new station in the area of former Fuchu Town.

As for bus corridors with frequent service, the plan is to subsidize vehicle upgrades and bus stops improvements to promote barrier-free access, not to subsidize service provision itself. The city government intends to retain frequent services indirectly through improving the physical environment for bus users and encouraging bus use.

10.3.2 Regulating Developments and Guiding Population Distribution

The basic land use policy is to focus on guiding land use, not reinforcing land use regulation. The city government intends to provide options of residential place and lifestyle and to realize the compact urban form through autonomous residential choices by citizens. Behind this is the recognition that development pressure is no longer intense and that land use regulation is less effective to shape the city under the circumstance of population decrease.

Regulation is strengthened on large-scale developments attracting many visitors, in accordance with central government policy[2], and sporadic residential developments in loosely regulated suburbs, the specific regulation method of which will be discussed around 2008. The city government, however, will not reinforce regulations on other kinds of developments. That is, it will not take measures to restrain housing developments which are distant from corridors, or allow higher building density in 'dumplings.'

On the other hand, an essential part of measures for guiding developments and population distribution is to create a more attractive living environment in the central area and along corridors by vitalizating public transportation (described above) and invigorating the city center (below). Another is to encourage living there more directly by providing financial and other support. The following two measures have been introduced.

The first is to subsidize the construction, acquisition and rental of homes which lie there and meet the prescribed physical and environmental guidelines. The support for central dwellings, along with the subsidy for provision of retail, medical and welfare facilities in lower-floor apartments, began in 2005 and been applied to 144 dwelling units over 23 months, which is nearly equivalent to the half of the newly built houses. It may have contributed to curbing the downward population trend. Support for dwellings along public transportation corridors, which has just began in 2007, is available within 500 m from rail/tram stops and 300 m from bus stops of frequent bus lines (Fig. 10-3).

The second is to support elderly households to move to the central area. Under the support program since 2007, the city leases a suburban detached house from an elderly household who wants to live in the center and sublets it to a household with small children looking for a suburban house, giving housing aid to the latter.

The city has set population density targets: 65 persons/hectare (2014) compared with 56p/ha (2004) in the central area, 50p/ha (2024) compared with 46p/ha (2004) in 'dumplings' along rail/tram corridors, and 40p/ha

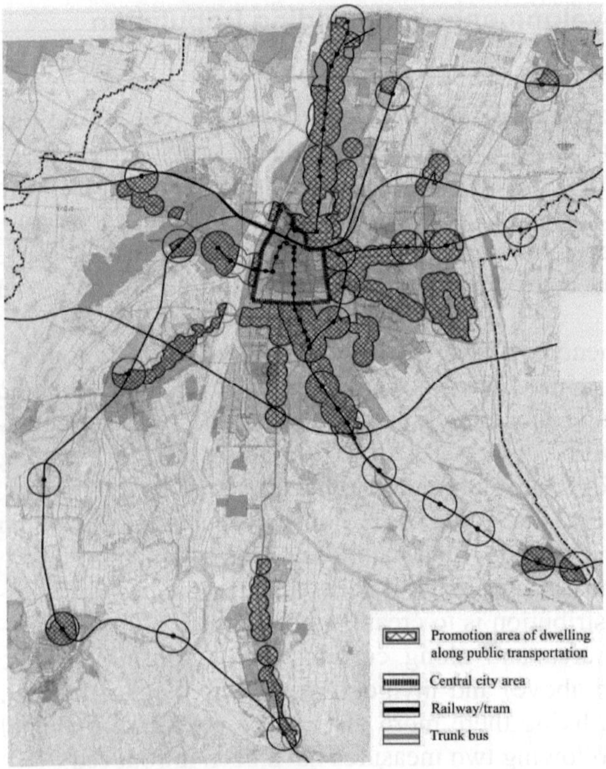

Fig. 10-3. Plan map for promotion of dwellings along public transportation (Source: Toyama City Government (2007d))

(2024) compared with 34p/ha (2004) along frequent bus corridors. These are not based on definite grounds but should be seen as a declaration of intention to increase density. However, the latter two seem to be met roughly by infilling vacant and farmland sites with average-sized detached housings[3]. The city government also aims to increase the ratio of those dwelling in areas well served by public transportation from 28% (2005) to 42% (2025) through guiding population distribution as well as vitalizing public transportation.

10.3.3 Invigoration of the City Center

As noted earlier, Toyama City obtained the first authorization of the second City Center Invigoration Plan (CCIP) from the central government in February 2007. Under this framework the central government is to provide intensive financial support to projects included in CCIP for the plan period

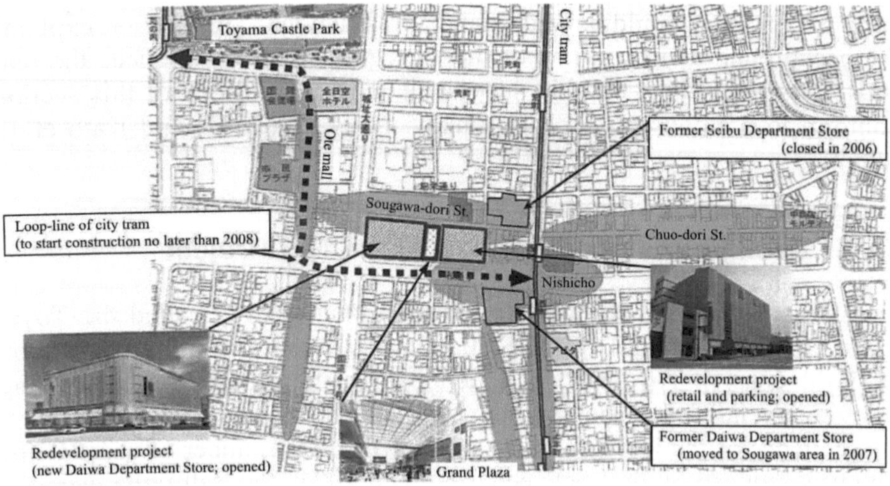

Fig. 10-4. Key projects to develop retail core in the city center (Source: Document provided by Toyama City Government in the interview survey (2007))

through 2011

Toyama's CCIP sets three themes: 1) making public transportation more convenient, 2) promoting residence in the central area, and 3) creating a lively city core. It sets numeric targets for each theme and has 27 projects, some of which were already described, in the 436-hectare plan area (Table 10-1).

Key projects expected to have a significant impact are the introduction of a loop-line tram, a retail redevelopment in the Sougawa area, to which Daiwa Department Store moved from a decrepit building in the Nishicho area, and the opening of the Grand Plaza next to it (Fig. 10-4). The sites of two former department stores, where mixed-use developments including retail and residential are envisaged as posterior uses, have been designated as special deregulation zones where procedures for opening new major retail stores are simplified in order to attract them. The city government also provides support for redevelopment projects from the consensus-building stage.

10.4 Toyama Light Rail: Fruition of the First LRT in Japan

As mentioned above, Toyama Light Rail (TLR) has been in operation as the first full-fledged LRT system in Japan since April 2006. In Europe, the LRT system has already established a firm position. However, it has been said difficult to put the LRT system into practice in Japan due to the difficulty in consensus building and financial restrictions. This section explains the background and overview of the TLR project and introduces the reasons why this project became successful. The conclusions of this section conform to previous research by Fukayama et al. (2007) and Kato et al. (2007).

10.4.1 Background and Overview of the TLR Project

In Toyama City, there was an eight-km-long railway called the Toyamako Line connecting Toyama Station and Iwasehama Station, located on the north coast of Toyama City. The railway had been used for commuting by commercial and business users and students as well. However, as a consequence of the increase in car dependency, the number of users of the line was reduced by half between 1990 and 2004. Naturally, the decrease of users caused revenue decrease and the operator, JR, had trouble maintaining the service level. At the same time, the Hokuriku Shinkansen is supposed to travel between Tokyo and Toyama in about two hours starting in 2014. As part of this big construction, a railway elevation program was instituted for the Toyama Station vicinity in 2003. Dealing with the Toyamako Line at Toyama Station was a serious problem. Elevating the Toyamako Line as well would be costly in comparison to the benefit. Although abolition of the line was also considered, a plan to continue the line by changing it into a LRT system was suggested as well. After deliberations by Toyama City, they decided to continue the line by changing Toyamako Line into LRT and to operate it by establishing a quasi-governmental corporation called the Toyama Light Rail Corporation.

Specifically, a section of the Toyamako Line rail track was removed for 1.5 km from Toyama Station, and in its place a new 1.1 km-long railway was built on the road as a tramway. The remaining 6.5-km section was reformed to meet the specification for the tram. Moreover, various serviceimprovements were undertaken at the same time, such as improvement in service frequency, extension of the last-train schedule, installation of barrier-free platforms, introduction of new well-designed low-floor light rail vehicles (Fig. 10-5), rail track greening, and so on. The total project cost was 5.80 billion yen, where 3.95 billion yen was for rail infrastructure improvement and the remaining 1.85 billion yen was for light rail vehicles. From a financial viewpoint, 2.30 billion yen was covered by the central government, 0.90 billion by the Toyama Prefectural government and 1.3 billion by the Toyama City government. Although the remaining 1.3 billion yen was supposed to be borne by the operator, in this case it was also covered by Toyama City as a subsidy. Note that the actual payment by

Fig. 10-5. Light rail vehicle used in TLR

Toyama City became just 0.3 billion yen since JR, the previous operator, provided 1 billion yen as a donation for regional development. For maintenance cost and renewal cost of the infrastructure and the vehicles, TLR Corp. can receive support from Toyama City. On the other hand, it is determined that labor and operational costs, including power costs, should be covered by fare revenue without any support from the municipality, even if there is deficit.

10.4.2 Success Factors of the Toyama Light Rail

Since it was the first LRT project introduced in Japan, there were naturally some barriers to accomplishing the project. Several important factors that led this project to success can be identified from various viewpoints. This part explains the major factors, which can be divided into two categories: pre-conditional factors such as technology and geography, and process management factors.

Pre-conditional Factors

- Existence of the Shinkansen project and railway elevation program: The first of the pre-conditional factors is the national-level decision in 2001 approving the Shinkansen project in Toyama. After this approval, various transport projects were found necessary in the Toyama Station vicinity, such as the railway elevation program. As a result, the budget of the railway elevation program covered most of the financial resources.

- Existence of railway infrastructure not used efficiently: The next factor is the existence of the Toyomako Line rail infrastructure, which was not being used efficiently. Thus, the discussion for LRT introduction could proceed more easily and smoothly because it would be easy to secure rail infrastructure for LRT with less new railway construction.
- Nonnegative financial balance condition: The third factor is the nonnegative financial balance condition. In Toyama, the population tends to decrease gradually and the number of people using public transport has been dropping because of high car dependency. On the other hand, a certain amount of residences and offices have been located along the railway that can generate travel demands at a certain level, and also there exists an urban planning project that will construct a huge attracting facility in Toyama City's central district. Therefore, cost-effectiveness analysis found that in the future the financial balance would be breakeven if the initial cost is supported by the public.

Process Management Factors

- Appropriate vision setting and strong leadership from the mayor: Toyama City promoted consensus building among stakeholders by regarding this LRT project as a first step toward 'compact city' development. The person who insisted that this is the appropriate vision was none other than the mayor himself. The mayor appealed to various local town meetings for the importance of this project and made efforts to persuade citizens. Therefore, it is safe to say that strong leadership has been necessary for consensus building.
- Holding various meetings and committees and disclosure of information: Toyama City held lots of meetings, from formal to informal, to discuss the LRT project at various steps of the policy formulation process. In particular, holding semi-formal meetings where mainly experts and other authority members discussed the issue is thought to be a contributing factor that gave rise to an efficient policy making process. Furthermore, all of the discussion points and conclusions were provided as news to the public, whether positive or negative. Toyama City also showed the LRT operation's deficit situation in earnest to ask local citizens to support the LRT system.
- Forward-looking stance to convert various constraint into positive aspects: For this LRT project, a good amount of financial support was provided from the railway elevation program. However, this program was not supposed to be applied to the LRT project. In addition to the

positive stance of the central government, Toyama City also lobbied the central government to improve the existing program. Consequently, the relaxation of the government financing schemes finally enabled Toyama City to bring the LRT project to fruition.

- Provision of incentives for the involvement of various experts: Since this LRT project was regarded as the first model case by the central government, it was also a chance for experts, engineers and other authorities to be involved. Due to this incentive, Toyama City could successfully involve lots of leading persons from various fields to exploit their useful knowledge and experience.
- The framing arrangement: Generally, people's decisions tend to differ based on the context, even if the target is the same. One of the reasons why it is difficult to build a consensus among stakeholders is thought to be the difference of common viewpoints, or 'framing.' And only if every stakeholder can share this common framing, discussion among them can proceed in the right direction and it becomes possible to consider countermeasures head-on against each stakeholder's concern. In this sense, Toyama City could successfully control and manage the stakeholders' framing by utilizing the policy or discussion authorized in the formal committee and obtaining opinions from experts outside Toyama City. Consequently, Toyama City successfully obtained strong support from members of the local communities and the city assembly as well.

Toyama City was able to successfully introduce the LRT system through combining all of the above-mentioned factors. Although the pre-conditional factors cannot be controlled, the process management factors can be achieved through local government effort.

10.5 Issues on Compact City Policy and its Framework

10.5.1 How the City Government Engages in Public Transportation

In principle, public transportation operators in Japan are required to provide services on a self-paying basis, and municipalities are not generally involved in planning and operation. Such a framework raises questions about how the Toyama City government should engage in public transportation planning and operation from the perspective of its compact city policy.

Firstly, public transportation 'sticks' were defined only as all of the rail/tram lines and existing frequent bus lines, independently of city planning issues such as the potential for accommodating population along the routes. The validity of this definition has not been explained clearly.

Secondly, the significant downward trend of bus ridership and frequency, which is often called a 'negative spiral,' suggests that even if a line currently runs frequently and is regarded as a corridor, it may become unable to maintain adequate frequency in the near future. To what extent the encouragement of living along corridors can contribute to ridership and service frequency is uncertain, as the density target is not set from the point of view of ensuring sufficient demand for public transportation. The city government has no plan to provide direct financial support to maintain service frequency at present, while it will give indirect support for upgrading facilities as mentioned above. It is surely difficult to guarantee direct support, which may discourage the bus operator from improving their ridership. Uncertainty about the level of service in the future, however, may make those who demand public transportation services hesitate to live along corridors. The city will have to find a way to deal with this possible dilemma.

These issues imply the inappropriateness of the present framework of public transportation planning and operation when it comes to implementing compact city policy. Closer coordination and cooperation with public transportation operators may be necessary.

10.5.2 Reasonableness of Not Regulating Urban Development Positively

It is often said that it is difficult to put strong regulations on land use in Japan, partly because that is seen as a violation of property rights. Given such circumstances and the city's recognition on land use control, emphasis on guiding measures may be realistic and reasonable. But it should be noted that many people in Toyama tend to believe it natural to build one's own detached house in the suburbs, at least so far. Therefore, it is considerably doubtful whether the relocation of population will proceed as planned just by deferring to citizens' preferences in residential choice.

Also unclear is how to develop the city center and 'dumplings' as walkable communities where one can meet the basic needs for daily life. Density targets in those areas were determined without relation to the population needed for day-to-day facilities to be economically possible. The city plans to consider a way to control and guide such facilities on the basis of interview surveys with businesses regarding intent of, conditions for and impediments to locating in the central area.

10.5.3 Relationship with the Prefecture and Neighboring Municipalities

Contrary to the city government's explanation that development pressure is not strong, Funahashi Village and former Fuchu Town (Fig. 10-1), both of which are located about 15 minutes from Toyama Station by rail, have recorded fast growth. Population of the former almost doubled and the latter increased by about 30% from 1990 to 2005. In particular, Funahashi Village can be an outlet for development pressure, because it is in a city planning area separate from Toyama City and is loosely regulated.

However, the city government sees these phenomena as being unproblematic because new developments in Fuchu area are almost completed and those in Funahashi Village are very few in number. As for the coordination of developments, it has not held discussions with neighboring municipalities so far and thinks it impossible to overcome conflicting interests to reach an agreement with them.

The policy for inter-municipality coordination of developments, which is the role of the prefectural government, is to be discussed under an advisory committee by 2008. In an interview survey we conducted, prefectural officials explained the position of the prefecture as follows: It collaborated with Toyama City where necessary, including contribution to TLR, but it was difficult to push compact city policy or show its intention about delineation of growth boundary, as it must pay attention to all municipalities in the prefecture. Although natural, the difference in motivation between the city and the prefecture is obvious. It will be a challenge for the city to construct cooperative relationships regarding the whole compact city policy with the prefecture and neighboring municipalities.

10.5.4 Possibility of Invigorating the City Center as a Retail Core

Unless there are attractive spots in the city center which can be reached easily by public transportation, neither public transportation use nor dwellings along corridors will increase as expected. In that sense, the most important element in the city center will be the retail function.

At one time, some retailers and landowners voluntarily tried several activities, such as shopowner incubation, called 'challenge shops,'[4] and the

City Center Thanksgiving Day, and achieved some positive results. However, some point out that movements toward invigoration by local businesses are generally not as active as in Kanazawa because local retailers and landowners tend to be inactive and dependent on the government. They also seem to be struggling to attract retailers from outside the prefecture.

The city government has only a vague vision regarding the central shopping area, in order not to constrain realistic urban developments as far as they are consistent with the direction the city envisages. Nevertheless, it has made efforts to guide them desirably for the city, including support for the start up of redevelopments and rebuildings in appropriate areas by dangling a carrot in the form of grants. It also hopes to stimulate private investments in the city center through measures including intensive public investments and deregulation under the CCIP framework, though the effect is still unclear.

10.6 Conclusion: From the Perspective of Local Governance

It is worth repeating that the mayor has played a prominent role in leading these efforts. He and his staff gave a comprehensible explanation about the policy necessity and the city's vision to citizens in town meetings and other forums. They position TLR as a leading project to evolve the compact city policy throughout the city. It will be meaningful to demonstrate the usefulness and smartness of a brand-new public transportation system to citizens in this car-dependent city. Also, the mayor and his staff do not necessarily emphasize short-term merit, and they often throw out questions such as "What if you get too old to drive?" and "How will your wife live a life without a car after you are gone?", which appeal to many citizens.

The city government, however, has not gone into issues which will be important but arouse much controversy, such as reinforcement of land use regulation and coordination of developments with nearby municipalities. In addition, the city successfully assembles various grants from the central government and invests them with a less financial burden for the city. These presumably lessen opposition from citizens, whether good or bad.

Another concern is citizens' generally low awareness of community design. Many people broadly tend to think it should be done solely by the government and are reluctant to participate. Even the public transportation operator seems to be unenthusiastic, as it has experienced dwindling ridership. Some hopeful movements have come up in the area along TLR since before its opening, such as mascot characters designed by nearby high school students, donations from citizens and local businesses, and tourism promotion. But, as a whole, it can be said that the transition toward bottom-up governance has made little progress.

At any rate, the step Toyama City has taken is significant for Japanese urban policy, as well as for citizens themselves. Within the policy framework outlined above, the realization of the city's vision is now up to whether citizens understand the significance of it, agree on it and act for it. The city is therefore required to maintain their support and, if necessary, seek a more effective framework in order to bring the vision to fruition, working more closely with other bodies.

Notes

[1] Urban Employment Area, proposed by Kanemoto et al. (2002).

[2] Major retail and other developments with floorspace over 10,000 m2 which will attract many shoppers and visitors are allowed only in commercial zones, neighborhood commercial zones and light industrial zones under the revised City Planning Law of 2006. In addition, restricting those developments locating in light industrial zones is required as a condition of authorization of the Central City Invigoration Plan.

[3] This reflects the city government's view that it is undesirable to reduce lot size and increase density of individual developments in order to preserve the environment of low-density residential area.

[4] Under the 'challenge shop' scheme, which originated in Toyama and spread to many cities in the country, vacant stores in the city center are sublet to young aspiring shopowners at lower rent for a limited period. More than 50 shopowners became independent and opened their own shops in the city center.

References

Fukayama, T., Kato, H. and Shiroyama, H. (2007) "Why has a new LRT system been introduced successfully in Toyama City?: An analysis from a viewpoint of policy process management", *Transport Policy Studies' Review*, 10(1) (in Japanese)

Kanemoto, Y. and Tokuoka, K. (2002) "Proposal for the Standards of Metropolitan Areas of Japan", *Journal of Applied Regional Science*, 7 (in Japanese)

Kasahara, T. (2007) "Effort toward Realization of Compact City Toyama", *City Planning*, 266 (in Japanese)

Kato, H., Shiroyama, H. and Fukayama, T. (2007) "Process Management in Public Transit Planning: A Case Study of the Light Rail Transit Introduction Project in Toyama, Japan" (Working Paper)

National Institute of Population and Social Security Research (2003) *Future Population Projections of Municipalities in Japan* (in Japanese)

Acknowledgements

Authors and members of cSUR research group on City Regions conducted interview surveys at the Toyama City Government, Toyama Prefectural Government, the Toyama Chamber of Commerce and Industry, and Toyama Chiho Railway, Inc. We would like to express our gratitude to all those who cooperated the surveys.

Part IV

DENMARK
GERMANY
FRANCE

11. Copenhagen, Denmark: Urban Regeneration at Economic and Social Sustainability

Hans Thor Andersen

11.1 Introduction

Copenhagen is the capital of Denmark, a small country in northern Europe with only 5.5 million inhabitants. The metropolitan region has about 1.8 million inhabitants while the central borough, the city, contains half a million. This gives Copenhagen a strong position as both the largest city in the country (number two, Århus, has a population of only 300,000) and the capital with all the associated functions: national government, the parliament, central administration, and major national organizations (industrial unions, cultural institutions, media, entertainment, sport and much else) which have, with one or two exceptions, established their headquarters in the city. Thus, Copenhagen plays a particular role due to its huge size compared to the rest of the country. As the biggest city, most legislation and initiatives in relation to urban conditions have had their start in the

Fig. 11-1. Denmark and Greater Copenhagen (metropolitan region)

city. There is a clear Copenhagen mark on urban thinking, planning and policies in the country.

As a residential city, Copenhagen has for centuries benefited from the presence of government, military and navy. During the period of absolutist monarchy, the city acquired a number of privileges to support its economic performance and in turn create an appropriate scene for the king and court. As a consequence, Copenhagen benefited strongly from this relationship and became the absolute dominating city by the end of the middle ages. During the 17th and 18th centuries, the profitable overseas trade was concentrated in the city and it became a trading center for northern Europe due to its brilliant position at the entrance to the Baltic Sea. During the late 18th century, the first examples of industrialization were seen and together with the highly profitable trade with China and India, Copenhagen experienced a golden period, which eventually was destroyed by the Napoleonic wars. Nearly half of all urban dwellers lived in Copenhagen at the time. The dominant position in the national context follows from the fact that Copenhagen had 100,000 inhabitants, while the largest provincial cities, Odense and Helsingør, had only 5,800 and 5,200 inhabitants respectively.

The 19th century completely changed the national and urban conditions. First of all, the kingdom lost its German provinces during two wars in the middle of the century (1848-52 and 1864). Nationalism was and is a very strong force; however, another outcome was the constitution of 1848. Absolutism was abolished and an elected parliament formed a free constitution, and what are now considered to be basic human rights were introduced. The result was a remarkable strengthening of civil society and the formation of national political parties. A further major change began in the late 18th century with reforms in the agricultural sector. The peasants, many of whom had copyhold, established themselves as a stabilizing social group when they became owners. In this sense, they formed the basis for the democracy that followed in 1848. Finally, the transformative processes of industrialization and urbanization accelerated during the second half of the 19th century and Denmark transformed from a typical farming country into an industrialized nation with quickly growing cities.

The industrialization soon put its mark on Copenhagen, as the city grew from 144,000 to 358,000 inhabitants between 1850 and 1900. By the end of World War II, the City itself had nearly 750,000 inhabitants, while the suburban areas contained some 430,000 inhabitants. Moreover, the growth seemed to continue. To control this growth, a group of planners produced the now famous Fingerplan, cf. next section, which has been followed by several newer generations of regional plans. However, the main ideas of the Fingerplan are still alive and an extended version was decided on August 1, 2007 as the basis for future planning in metropolitan Copenhagen.

Copenhagen was the national center of the manufacturing industry to the late 1950s, and nearly half of the national employment was found in central Copenhagen. However, the strong expansion of welfare services and industrial restructuring turned the capital into a city of services, both public and private. Yet, the economic turbulence of the 1970s introduced a long period with substantial difficulties. The manufacturing sector accelerated its decline and by the turn of the century, employment has dropped from nearly 200,000 to less than 5,000 jobs in central Copenhagen. A few factories moved to suburban or provincial locations, but most simply closed. As a consequence, the unemployment level rose from 1-2% during the 1960s and early 1970s to 15% and higher in the 1980s. From 1960 to 1990, the population of the central city declined 721,000 to 466,000 inhabitants; however, the out migrants were the well-paid, middle-class families that left the single, low-income groups and pensioners behind. To finance basic services, the city had to obtain loans and thus debt grew quickly. The economic well being of the central city was highly problematic and the city went technically bankrupt in the early 1990s. A part of the problem was the national policy of decentralization, which transferred major investments in infrastructure, health and higher education to the provinces. Copenhagen suffered at the same time from decay of large sections of former industrial areas, in the harbor and rail land, as well as an outdated stock of dwellings from the late 19th century.

The depressing situation with high unemployment, population decline, a growing debt and a number of unsolved problems forced the national government to react. One major reason was the European dimension.

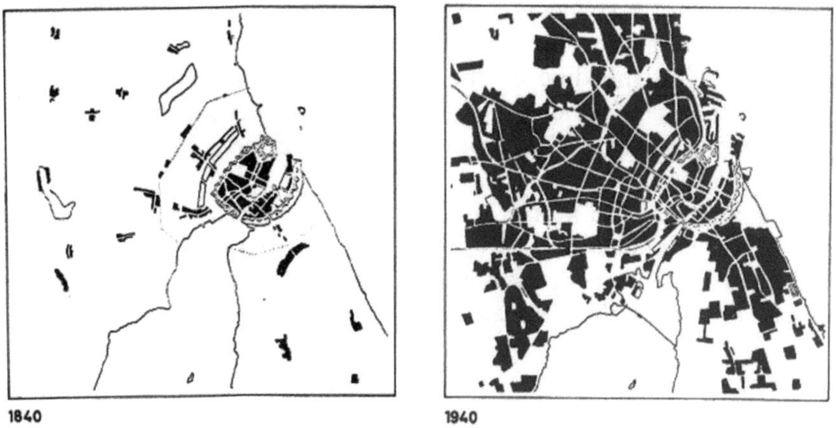

Fig. 11-2. Built-up land in Central Copenhagen 1840 just before the fortification was abolished and 1940 before suburbanization began. Each map covers 10 x 10 km. (Source: Hovedstadsreformudvalgets sekretariat (1970))

The emergence of the Single Market was expected to have major impact on larger cities. They would loose their prominent position at the top of a national hierarchy and instead become parts of a European system of cities. Those who could attract important functions in government, finance, insurance and headquarters of large businesses would be the winners in the European territorial competition. Moreover, the industrial restructuring gave high-tech, research and development as well as cultural functions a clearly more important position. Thus, big cities, not remote rural landscapes, were then considered to be the national assets in a Europe of free trade and movement. The government realized that the only real big city was the capital. In quite short time, Copenhagen shifted from being a problem to becoming a key platform for future welfare and the government called for an open debate on March 20, 1990. This debate triggered a round of major focus on the city as well as a long list of central government initiatives, cf. below.

In the sense that population and employment growth are markers of the state of a city, Copenhagen made remarkable improvement in less than five years. Suddenly, the city has received a lot of private as well as public investments, the property market is heated, most employment sectors lack people, unemployment is decreasing year after year and the general income level is growing. The initial investments by the national government have triggered a major and it seems lasting amount of private investment in the city. Formerly derelict land is now either turned into housing or offices or is about to be redeveloped, and the transformation of the old harbor and city annex of Ørestad has been most successful.

This chapter aims at clarifying the framework for a fundamental shift in urban policy from a social justice approach to an economically and socially sustainable approach. The latter is addressing a lasting balance between income and spending by the local government, a balance basic to any attempt to regain some degree of political independence. This has put regeneration at the heart of urban policies in the effort to achieve this balance.

The paper starts with a brief outline of the Copenhagen region and its economic and administrative development in order to present the framework for the actual regeneration policy. The main effect of the structural crisis and failure of the administrative system became clear in the late 1980s and created a drastic change in urban policy during the 1990s. Strategic planning became a key issue. The planning system and its relationship to the urban strategic policy is analyzed in the following section. The most important shift is the strong partnership between central government and the City of Copenhagen, which has determined regeneration policies since the 1990s. The regeneration policy and its implementation is the focus in the last section. The materialization of entrepreneurial strategy has become visible in former industrial and harbor areas.

11.2 Metropolitan Copenhagen: An Outline

The administrative arrangement of Denmark was for centuries organized around about 1,200 perishes plus 25 counties and nearly 80 cities with certain privileges in relation to trade and industry. The first wave of modern urbanization (1850-1920s) could find room inside existing city borders, while the second wave after World War II clearly crossed most existing administrative borders and called for a reform. This reform came in 1970 and consisted of several elements: first, a guiding principle was to include de facto connected parts of an urban area in one administrative unit ('one city - one municipality'); second, to decentralize major parts of central government competence as well as financial resources to the local (municipal) level; third, to maintain the principle that those who decide also pay in order to maintain financial responsibility; fourth, that the local level must have a certain minimum size to obtain economic efficiency; and fifth, the division of work among municipalities, counties and the state should be clear and relate to the spatial dimension. All public services close to citizens (public aid, housing, schools, culture, and secondary roads) were relocated to municipalities, while more regional functions (hospitals, health care, public transport, secondary educations, primary roads) were in the hands of the counties.

The outcome of the reform was extensive, as the 1,200 perishes were reduced to 275 municipalities, and the cities disappeared as administrative units after they were connected with the surrounding perishes to form the new municipalities. Moreover, the 25 counties were reduced to 14, plus the central boroughs of the capital, namely Copenhagen and Frederiksberg. The larger municipalities combined with the strong decentralisation and the massive expansion of the welfare state during the 1970s had a tremendous impact on the regional balance. The provinces experienced strong improvements in employment, services and local decision-making power, while the capital region already had established most of the welfare services during the 1950s and 1960s.

The reform had a major drawback seen from the capital, as the principle of uniting de facto coherent urban areas into one administrative unit was not followed in the capital. This unwillingness had several causes, one being the resistance from the conservative party, which had and has its strongholds in several suburban municipalities. Another is the fact that a united Copenhagen would form a municipality of 1.8 million inhabitants, which would increase distance between local politicians and the population. Besides, the sheer size would make one municipality have almost a third of the national population. To compensate for this, the government organized a fourth tier of government, the Greater Copenhagen Council. Its main obligations were regional planning, medical services, public transport and transport infrastructure (Andersen, Hansen and Jørgensen (2003)). Moreover, some cultural functions with regional relevance were added. The council was made up by 25 appointed members, often local mayors. The council existed to the end of 1989 when a conservative government decided to follow the examples of the UK and abolished this extra

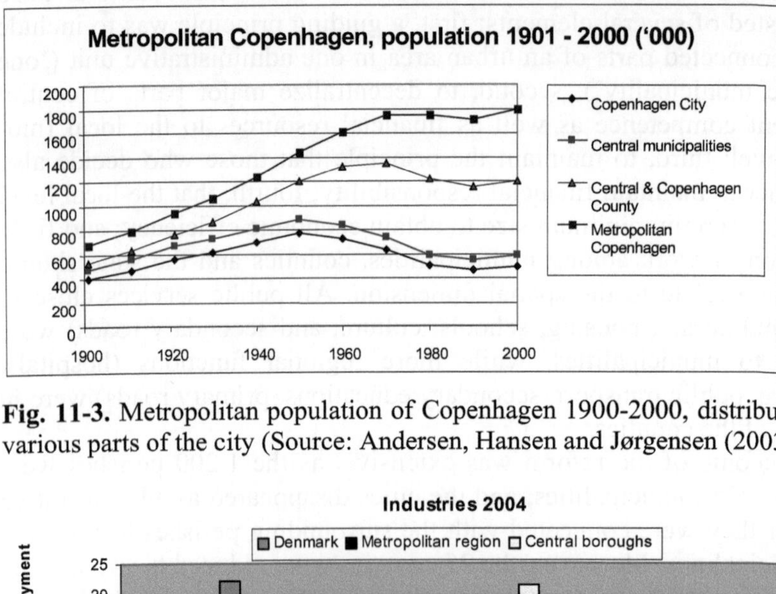

Fig. 11-3. Metropolitan population of Copenhagen 1900-2000, distributed among various parts of the city (Source: Andersen, Hansen and Jørgensen (2003))

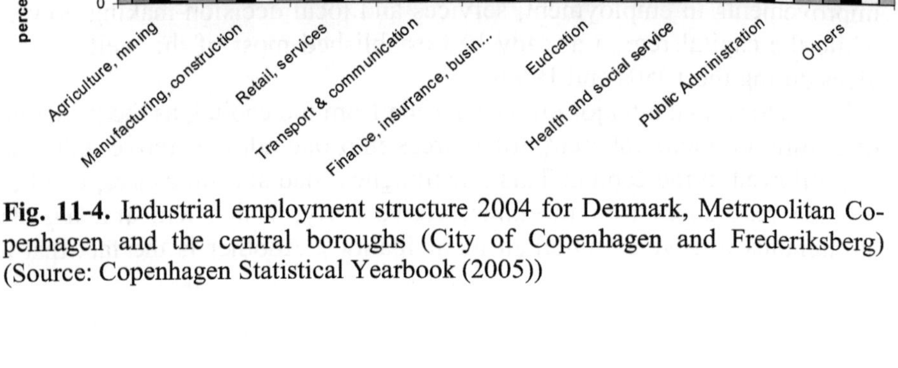

Fig. 11-4. Industrial employment structure 2004 for Denmark, Metropolitan Copenhagen and the central boroughs (City of Copenhagen and Frederiksberg) (Source: Copenhagen Statistical Yearbook (2005))

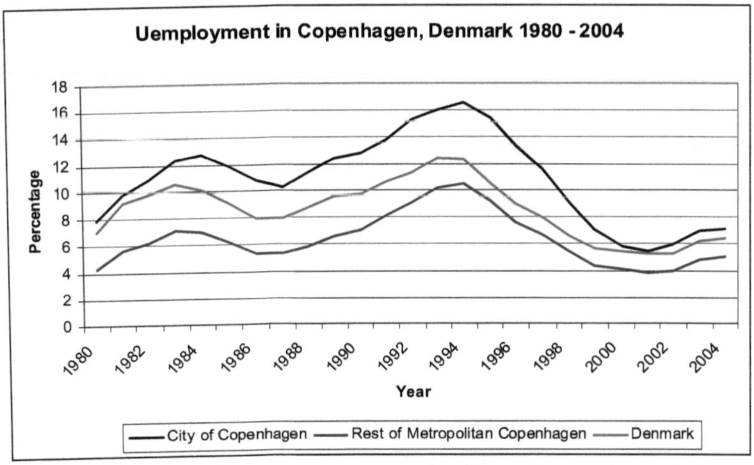

Fig. 11-5. Unemployment registrated (percentage) 1980-2004 for Copenhagen, Metropolitan region (Copenhagen excluded) and Denmark (Source: Statistical Office, Copenhagen City)

tier of government. The coordinating functions were transferred to central government offices and the counties, except for public traffic, which was organized as a quango - HT (Copenhagen Public Transport).

The Copenhagen region peaked as a manufacturing region in the 1950s. Since then employment has declined year by year. However, during the 1960s and early 1970s, the expansion of the public sector compensated for the job loss in manufacturing; yet, as the financial crisis of the state forced governments to reduce public spending, unemployment increased from the late 1970s and until the early 1990s (Fig. 11-5). A new financial policy from 1993, when government changed, fuelled the economy, and combined with a targeted urban policy, things improved quickly during the 1990s. The city has managed to create a new industrial base in finance, insurance and business services, which is able to generate general economic growth in the region and reduce unemployment.

Urban regeneration has of course always existed as an unplanned and uncoordinated activity. Derelict buildings have been replaced by new ones as need demanded it and resources were available. Understood in a more modern sense, urban regeneration began prior to World War II with a strong focus on run-down neighborhoods built in the mid 18th century, and the former main station. The former, which were quite centrally located, were cleared over a five-to-ten-year period and then rebuilt as modernresidential quarters. The demand for regeneration was based in the very low housing standard, low maintenance level and concentration of severe social problems. The latter was a highly centralized space close to the

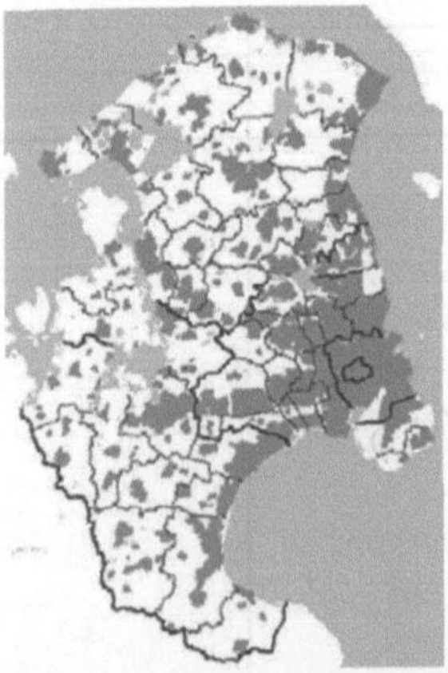

Fig. 11-6. Metropolitan Copenhagen 2004; urban built-up areas and administrative borders (Source: Statistical Yearbook for the Copenhagen region (2004))

Fig. 11-7. Four typical forms of housing in central Copenhagen: Upper left: Private renting from the 1910s, upper right: non-profit housing from the 1940s, lower left: detached housing from the 1920s, lower right: non-profit housing from the 1950s

existing central station and the Town Hall. The area was vacant after a major reorganization of the city's rail system at the start of 20th century. Uncertainty about future use delayed reconstruction of the space, but finally, during the 1950s and 1960s, the whole area was turned into office space and a major hotel. The public sector co-financed the former project, but not the latter one, which was considered to be a commercial regeneration scheme on market conditions.

Urban regeneration was dominated by the concern for better housing to the 1980s. At that time, the wider urban environment was included in urban regeneration. But the perspective remained focused on welfare and housing. At this moment, the social problems of certain estates in the suburbs qualify these estates for concerted action. Their situation demonstrates that an above standard of physical conditions is insufficient to solve urban problems. Since then, a rising share of local improvement programs has been located in suburban areas, while inner cities have peaked. The last major improvement program that focused primarily upon housing conditions was the Vesterbro Neighbourhood Scheme. The scheme involved some 10,000 inhabitants in a former working class neighborhood located close to the CBD. The lack of maintenance and a dominant proportion of small dwellings combined with general lack of modern equipment had turned the district into an area populated by third world immigrants, students and many poor and unemployed people. After the renovation, the dwellings have become modern equipped and enlarged; as a consequence, the district has now become the hot spot of the city with the highest property prices. A conversion from private rentals to shared ownership or private ownership has changed the social composition of the district towards young academics and middle-class people with above average incomes.

The experience that a conversion of run-down private rentals could raise the social position of an area was soon taken into strategic consideration. Until the late 1990s, only non-profit ('social') housing companies were willing to construct new housing in the inner cities. The overall problem with the non-profit housing was the automatically increasing number of low-income groups, people non-active in the labor market, and immigrants from the third world. The outcome was higher social costs and less income from taxation. The only advantage was the social dimension; it held a door open for the poorest to the new housing.

The City of Copenhagen learned the lesson and determined to stop further non-profit housing units. Instead, a private housing scheme for sale was promoted with great success. A report from the strategic unit concluded that 13,000 new privately owned dwellings would give the city long-term economic sustainability. With such a number of new privately owned dwellings, young urban professionals could be attracted to the central city and raise the income level. The outcome of this was on the one hand a strategic search for spaces for the new dwellings, cf. following section on urban regeneration. On the other hand, the city shifted its priorities to a close partnership with the national government, as well as large, private investors. The main example was the Ørestad-project combined with the fixed link to Sweden. Both projects materialize the new will and strategy for competitive growth policy.

Over the last five-to-ten years, the aim of regeneration has not just been physical repair, but long-term improvement of the general economic and social climate of the city. The change has turned attention away from social justice in favor of economic competitiveness and as such represents a shift towards a more entrepreneurial policy style. While Copenhagen is not a global city, it strives to participate in a global context and this has put its mark on the policies adopted.

11.3 Strategic Planning and the Planning System

Planning in a modern sense arrived on a voluntary basis in the 1920s. The reason was the growth of Copenhagen, which called for coordination. The first act on planning passed parliament in 1925, but never had any effect. The crisis of the 1920s and 1930s reduced the need for planning legislation, but in 1938 a new act on urban planning was passed. The act demanded that all cities and urban localities with 1,000 or more inhabitants begin a planning procedure. The outcome was to consist of a general plan for the whole municipality to guide and coordinate future urban planning.

Secondly, the local government was to make action plans for specific developments. Few of these plans were made before the 1950s, and in particular, the City of Copenhagen took its time. In fact, the first municipal plan was not made before the early 1980s --- nearly fifty years after the act was passed. The Fingerplan was the outcome of a non-governmental group of urban planners, who were able to convince the government to make a study and produce some guidelines for future urban growth. The plan was highly successful, as it still forms the basis for planning.

The efficiency of the planning system was rather low, and in too many cases the local governments' interest in and understanding of the procedure was quite limited. Especially during the period of rapid suburbanization, the negative effects of these lacks became obvious. Real control of land development was needed, but unfortunately, the proposal failed in a public referendum in 1963. First by the end of the 1960s, a new act on national and regional planning was agreed upon by the parliament. Thereform of the planning system was linked to the municipal reform, cf above. The basic principle is that all local planning has to be in accordance with regional planning, which in turn must fit with the national planning priorities decided by the minister and the parliament. The minister presents the government's priorities every year, since the mid 1990s every fourth year. These priorities should be included in regional as well as local plans. Moreover, the decisions for planning have a binding effect for only public authorities and their priorities. Only the municipalities can produce plans which are compulsory for the citizens.

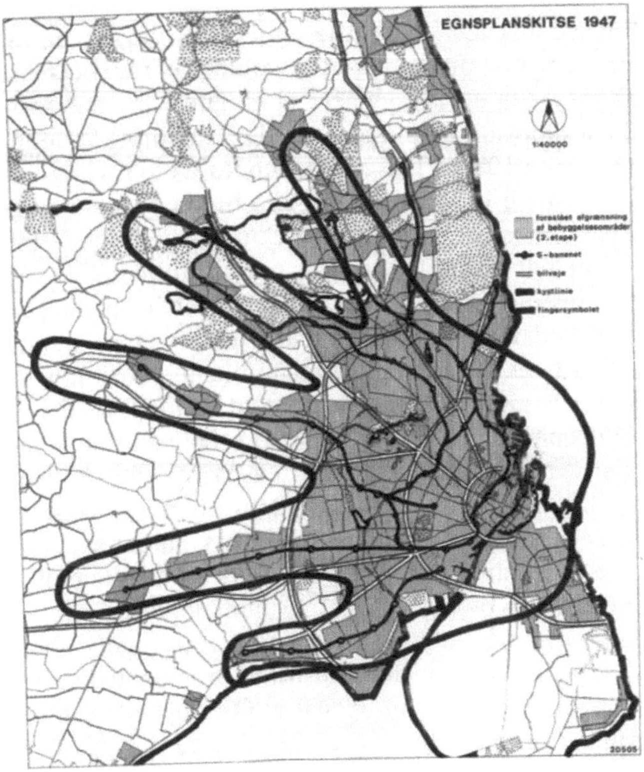

Fig. 11-8. The Fingerplan, original published 1948, was inspired by the Abercrombie's Greater London Plan from 1944 (Source: Engsplankontoret (1947))

The planning system fits within the government system --- national government produces guidelines, which every other tier of government is to follow, regional governments (counties) decide the regional structure and coordinate the needs for infrastructure developments, natural preservation and public transport through their regional plans, while the local governments (municipalities) make a municipal structure plan and a series of local plans to fill the overall ideas for development.

The governmental reform tried to relate responsibility to competence, i.e. that the government in charge should bear the costs as well as have the benefits of its own decisions. Unfortunately, this was never the case in Copenhagen. The administrative fragmentation and the system of income taxation made it tempting for many municipalities to 'free wheel' – if they focused upon upper-middle class families, they could create a strong, local economy as they then could combine high-income groups with little need for public support and low taxation. In contrast, working-class districts would face a difficult economy due to a large demand for public services and lower incomes to finance that service. At the same time, those who invested in industrial improvements, marketing etc. would have to bear the costs, while the benefits may end up everywhere else. Thus, the administrative system produced irresponsibility and paralyzed all attempts to counteract the structural crisis of the 1980s. The consequences were raising unemployment, a very slow growth in incomes and an obviously stagnating city. Yet, the problems were mostly felt in central Copenhagen, while the affluent suburbs hardly ever felt unemployment or social problems during this phase of economic restructuring. When the Greater Copenhagen Council (GCC) was abolished from 1990, the main functions --- public transport, economic development, environmental planning and hospitals --- were transferred to quangos and the central government. Copenhagen Capacity and Wonderful Copenhagen were set up to promote business investments and tourist visits, respectively. A main reason for the abolishment of GCC was its steadily increasing weakness. Controlled by a council of mayors, it had no chances to develop a strategy across the interests of the municipalities, and the appointment of the councillors eroded its political legitimacy. Moreover, the growing power of municipalities simply squeezed the intermediary --- the county --- level, and it was finally abolished at the end of 2006. This leaves Copenhagen with a fragmented structure of now 35 municipalities which individually develop their local strategies, but not coordinate them. Only to a smaller degree will the new region (which does not include the southwestern part of the capital region) be able to guarantee the abolished coordinative function.

The new local government reform reduced the number of municipalities to 98 and abolishes the counties from 2007. Only 5 regions will take care of coordinating a few regional issues, the first being public transport and hospital services. All other obligations are transferred to municipalities. In relation to this, the planning legislation has been changed so that a regional plan is not needed anymore. Instead, municipalities will have to maintain the overall concerns and attempt to include the interests of the whole region.

The last few years have seen a remarkable shift in the climate for cooperation between the City and the central government. While often disagreeing in the past, the present situation shows mutual understanding and trust to ensure major initiatives. The decisive break with the past was the willingness to invest in public transport --- new regional and suburban trains, increasing capacity of the road system, massive investments in research, higher education and culture together with the Øresund link, a combined rail and motorway tunnel/bridge between Copenhagen and south Sweden. Finally, the decision to construct --- and co-finance --- the new metro broke the ice. During the late 1990s and later, the mutual interests of the state and the City have been obvious. The City and the wider Greater Copenhagen have benefited from the massive central government investments, but the statc gains from the informal partnership by steady economic growth and economic modernization, which improve the competitiveness of the country.

Much in contrast to other cities in the country, private business leaders rarely take part in local strategic policies, at least not in a visible way. While many large cities have been able to develop strong leadership capacity thanks to strong cooperation among local government, business leaders and relevant unions and other organizations, this has never really been the case in the capital. At the darkest moments in the late 1980s, no initiative from the business community was delivered. It was hardly possible to convince the business people to participate in attempts to generate new knowledge-based developments. One reason of course is the administrative fragmentation, and another reason is the city's position as national capital. As a result, business leaders consider themselves to be nationally or internationally oriented, not locally focused. In this sense, the City relies upon the ability to cooperate with the central government, which is possible, but always at the price of central government interference in local matters.

The general climate for strategic development and business promotion has been rather poor for a long period; yet, the main task seems to end up in the City of Copenhagen (which has only a third of the population in the region). The suburban municipalities have constantly preferred to free wheel, including those with the highest shares of unemployment. This gives the central city a strong position as the political decision-making center of the region, just as the close cooperation with central government has strengthened this position. The experiences from the last fifty years are twofold. Firstly, only when the City of Copenhagen takes the leadership is it possible to formulate and implement a strategic plan for the whole region. The relatively strong suburban municipalities have no competences and do not feel responsible for anything other than pure local interests. Secondly, a mutual understanding with the central government is needed to carry through major projects with a structural effect, i.e. projects which have the potential to change the course of the city or make a decisive improvement in businesses. This last item points to the City's dependency on national government policies. The 1970s and 1980s were years relatively hostile to Copenhagen, as decentralization and equal development were the basic principles of national policy. The remarkable shift in national priorities has re-established the close relationship between national and regional policy in the capital.

11.4 Urban Regeneration

The urban development in Copenhagen took a decisive turn around 1990. Throughout the 1980s, the political situation at the national, regional and local levels remained in the tracks of the previous period. Thus, decentralization was the main theme of national and regional policies, while local politics was divided between a need for modernization on the one hand and attention to social justice in urban renewal on the other. The ideological contradictions were personalized through the lord mayor and his primary political opponent, the left socialist alderman for local planning. Through a decade, the right and left wing managed to stop any major initiative due to their mutual contradictions. The conflict revolved around the future for Copenhagen as either a competitive, European city versus a comfortable welfare city (Harding (1997)). During this period, the city suffered from a still more severe financial situation, as the left-wing resistance to reduce public services in keeping with population decline contributed to a sharp increase in the debt of the city. At the same time, the national government did not invest in the capital due to the stated policy of regional equality. Moreover, economic restructuring produced rising unemployment, and severe social problems were visible in Central Copenhagen.

The combined effect of a new lord mayor and a central government concerned with the capital in response to the structural problems and the emerging European competition among big cities all accumulated in a cry for initiatives. Thus, in the late 1980s, the conservative prime minister set up a commission to produce new ideas and propose major improvements in the economic performance of the city. At the same time, the national government had started to formulate new priorities, and the parliament debate on the March 20, 1990 marks the official policy change. Shortly after, the government and the social democratic opposition made a major agreement which gave the Øresund as well as the Baltic region high priority in national objectives. Shortly after, Copenhagen was designated European Capital of Culture and the central government fueled the activities by expanding/rebuilding many existing museums and cultural institutions. Furthermore, the decision to construct a 20 km-long tunnel/bridge to southern Sweden and the Ørestad (city annex) connected to existing urban center by a new metro helped the city to regain former strength.

The most important change in this relation was perhaps the complete change of the political environment of the City of Copenhagen. The new lord mayor managed to provide an unprecedented general agreement in the local parliament and thus create a strong platform for negotiations with national political leaders. In the meantime, the national government had realized the impossibility of a regional initiative to recreate the dynamism and vitality of the capital. In addition to the competitive pressure from outside, i.e. Single Market and similar deregulations, the national government began to regard Copenhagen as a key asset in the evolving European Union. This new recognition implied also an abandonment of the 'technocratic'---hierarchical system of government. Instead, ideas imported from Anglo-Saxon countries were supported, namely close partnerships that short-circuited the usual line of command, often based in ad hoc projects financed in more or less spectacular ways, cf. the presentation of the Ørestad project below. While the 'classic' government mode of discussing and formulating new initiatives was a long, but open, process based on a number of compromises and more or less irrelevant considerations, the 'new' form of policymaking took place in discretion between a few trusted partners with the necessary resources mostly aiming at a specific project.

The overall mark of this change in policy is usual called a shift from 'government to governance'; it reflects the different climate of decision making as well as a turn to a clear proactive style of strategic policy (Harvey (1989)). While strategic planning before the 1980s most often consisted of 'enabling,' i.e. providing the framework for desirable developments, the last decades have seen the rise of a strategic policy style which focuses on 'completion.'

In the case of Copenhagen, the coincidence of several major changes gave the city excellent opportunities, which a new government recognized and was able to utilise. The shift in national government lead to a change in favor of a more expansive economic policy. Within a few years, employment grew quickly, investors began to focus on the central areas as a field of opportunities, and the general economic activity in the city took a clear upward turn. The conservative government made an agreement with the opposition for extensive investment in the infrastructure of Copenhagen (Andersen & Jørgensen (1995)):

- A tunnel/bridge across Øresund, the narrow water between Copenhagen and Malmø, the third-largest city in Sweden, including a motorway and rail line to the airport
- In relation to the fixed link, a number of government initiatives followed, including the fora and networks for cooperation across Øresund, spread over a long list of activities from universities, tourist promotion, labor-market cooperation, medical industry (medicon valley) and other high-tech industries.
- The government has decided to enlarge the capacity of the central rail station, as well as the west motorway from six to ten lanes, motoring

3 from four to six lanes, and finally to build a new north western motorway in order to improve the connections to the outer north western parts of the metropolitan region.

- The cooperation between the Danish and Swedish governments has put the region at a high priority in the European cross-border programme INTERREG --- Øresund and the Baltic region have both received major support from the EU.
- The decision to announce Copenhagen as the European City of Culture in 1996 was an occasion to modernize and expand the museums and national art galleries, royal archive and the royal library. A new opera house was planned and stopped before a private fund offered the city a new opera building, which opened 2005.
- The decision to remove the navy from its historical location in Copenhagen to a provincial town furthermore opened up new development opportunities right in the middle of the city, and exclusive housing, advanced business services, media advisers and a number of cultural institutions have been located in the area less than 0.5 km from the city center across the harbor.
- The government launched a new program for neighborhood improvement and backed the intensive Vesterbro improvement scheme. Both accelerated the needed upgrading of the old housing stock and triggered a series of private investments in the housing sector.
- The Ørestad and metro-project: A city annex stretching 0.5 km from Town Hall and linked to the existing central areas by a new metro line.

These central government initiatives all helped to improve the settings of the city, and the ability to benefit from this was provided by a new governance style in the Town Hall. The new lord mayor appeared to be an integrative politician who managed to create solid support for his policies across the political spectrum. Moreover, the positive situation the city got thanks to the investments listed above was further used to create a strategic development plan focusing on financial sustainability. With the new style of government and governance developed in the Town Hall during the second part of the 1990s, the City appeared to be a most proactive partner for the national government as well as for private firms, the European Commission and various European organizations.

A cornerstone in the new strategy was the recognition of the connection between housing policy, urban regeneration and the local fiscal situation on one hand, and on the other hand the political room to maneuver. A strategic task force at the Town Hall did several studies of housing policies and its fiscal effects and concluded that the then positive situation was rooted in general favorable economic relations, but that a slowing down of the economy in general would bring the city back to a difficult economic situation. The group also suggested a proactive strategy, advising that the city should plan for at least 13,000 new housing units for sale within a five-to-ten-year period. If it managed to attract the right kind of people, i.e. well-educated, middle-class families with good incomes, the economic balance of the city would tip in favor of long-term sustainability. The municipal strategy plan has carefully evaluated each major space for potential regeneration and calculated the public costs versus the short-term and long-term gains. On the basis of these calculations, the City has made a priority of regeneration projects.

The Town Hall formulated in 2001 a policy statement "Housing for All," which argued for "... an economically sustainable population." To reach this position, new houses had to be constructed on available space, both in existing residential areas as well as on hitherto industrial and harbor land. The City began to identify possible spaces for housing construction. The most obvious space was and is the harbor area. A study called "City by the Water" pointed at possible locations for new housing in the existing harbor (Fig. 11-9).

"City by the Water" identified four different areas of the Copenhagen Port realistic to develop: the former free harbor in the north, the inner harbor and navy area at the center, to the east a coast line with big potential, and to the south west a run down industrial harbor ready for redevelopment.

The successes from the first round have triggered further plans. The

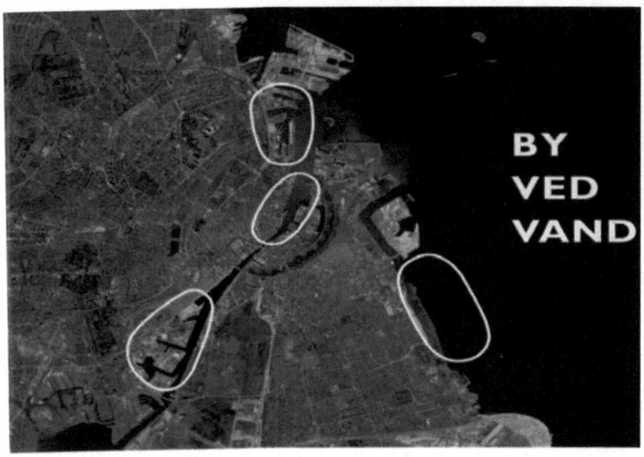

Fig. 11-9. "City by Water" – a strategic study for grounding the "Housing for All" plan of 2001.

Fig. 11-10. Ferry terminal, housing and offices in the former free harbour; most of the vision is already constructed (Source: By ved vand)

Fig. 11-11. Havnestaden, a mix of owner-occupied, shared-ownership and rented housing in a former industrial area close to the city center

redevelopment of Nordhavn (North Harbour) from today's storage and empty space to a residential area with office functions has just been decided (October 2005). The plan allows for 600,000 m^2 of office and housing space. The new urban area is planned to be linked to the existing city via a metro line.

Fig. 11-12. Rail land near central station, hitherto used for shunting, parking, cleaning and maintenance of wagons. Located between the inner city and the now refurbishes harbor, the space is highly attractive for redevelopment

Fig. 11-13. Inner-harbor redevelopment; offices and hotel (Marriott) constructed close to the central station

The most important and regenerative project decided by the City is the Ørestad project and the metro system (Andersen and Jørgensen, 1995). The Ørestad area is a 3-km² space located about 1 km from the Town Hall. The area was jointly owned by the state and the City of Copenhagen and formerly used by the military. The land was reclaimed from sea in the 1940s. The Ørestad project was formulated at the end of the 1980s by government officials in the department of finance, the national railways and the City. The principle idea was to develop a centrally located city annex with excellent accessibility for advanced services and research. The project would give the city a major boost without costing the national government anything. By developing the land, including metro line and motorway junctions, the land value would increase considerably and be able to finance the whole infrastructure development. Over the next 30 years the loans for this development could then be repaid. The metro is now in operation and highly popular, although the anticipated number of passengers, which shall pay for the services, has not been reached.

Fig. 11-14. At the University station/Ørestad looking north, August 2007. At the right the new media house of the national broadcasting company

Fig. 11-15. Ørestad station at the crossing of the rail line between central station and the airport/Malmø. On the left the newly opened shopping mall called Fields

Shortly before the end of 2005, the City of Copenhagen, the Borough of Frederiksberg and the national government agreed to construct and finance a major extension of the existing metro system, the so-called ring metro project. The costs are calculated at 15 billion Dkr (about 2 billion euro) and the project is expected to be in operation in 2017. The project will be financed by revenue from selling the city's power plants and selling land in the harbor. The revenue from running the metro system implies a major regeneration of parts of the harbor to provide the needed number of passengers. Thus, the decision to construct this ring line has major impacts on the strategic plans for the City, as it has to regenerate the harbor areas to provide the necessary funding.

The first metro line as well as the project for the ring line are both centerpieces in the municipal structure plan as well as the regional plan. However, the decisive partnership was the negotiations between the state and the City before the official decision making. The plans serve primarily as the official declaration for the future development of the City. Hence, they appear as the visible expression of the negotiations and discussions among various partners. Thc complete list of active participants in the decision-making process is unknown. Only the key actors such as the involved municipalities and the state, the port authorities, and the metro company are known.

Urban regeneration has for long been recognized as necessary to ensure the economic sustainability of the city; however, neither the pace nor the extent of the projects can be ignored in this relation. The City of Copenhagen calculated that area located adjacent to existing rail stations should have priority since they usually do not demand further infrastructure investments. On the other hand, if such investments are needed, then the speed of redevelopment must be high to pay back initial costs. The overall conclusion is that spaces with potential for regeneration have to be carefully chosen in order to maintain sustainable financial development.

Summary

The turn from 'old'-government-style policymaking to modern-governance-style does not imply that the existing planning system has lost its function or relevance; however, the hierarchical planning model exists side by side with the new 'direct'-project planning mode. The functions are strongly related to the political and the juridical relations, as the planning system is negotiated between elected authorities and passes a public examination before decision making. This in itself brings legitimacy to the planning system. The internal logic of the system, that all planning at lower tier has to be in accordance with the framework of the upper tiers, makes the planning system an efficient mediator of government decisions and priorities among various tiers of government as well as public organizations. The juridical function is obvious as the adopted local plans bind legal rights and burdens to the individual property. This implies that the planning system is not indifferent, despite the new governance form of decision making/implementation.

In the case of the Ørestad city project, the National Planning Agency produced a report ("Denmark towards the Year 2018") in the wake of the central government-municipality agreement which gave the Ørestad development a high priority as national spatial and investment planning. This was soon followed by the City's own municipal structure plan and the regional plan. The positive spirit from this agreement has further involved other quangos (the harbor authority, the property corporation called Freja, and various government organisations such as universities, national broadcast, and national archives). However, private investors have been surprisingly invisible in the debate and planning process. The funding of the project has from the beginning depended on central government and City resources. The trick to finance the infrastructure investments by the expected increasing land values due to the improved infrastructure has only been possible thanks to the central government guarantee. All deficits or refinancing is based upon the state. This is to say that the financial sustainability is based upon the taxpayers. Other projects have a broader engagement at the local scale, with citizens as well as private business; however, they rarely involve funding from the private sector.

Two important conclusions can be drawn from the last 20 years of urban strategic policy in the Copenhagen region. Firstly, that the attempt to create a regional authority which could formulate and implement long-term strategies clearly has failed. At least as long it consisted of a council formed by a number of mayors, they have proved unable to convert local interests to those of the wider region. Secondly, only when the City of Copenhagen manages to act proactively and cooperatively in a partnership with central government has it been possible to make policies with a horizon beyond daily operations. Such a period arrived in the 1990s, perhaps only because the City was in severe financial difficulties.

Copenhagen has shifted from a cold to a hot spot in ten years' time, but the success demanded a major shift in priorities – from comfortable welfarism to a European competitive city. The turn to entrepreneurialist policy style has had a price in terms of a reduction in public engagement and shift in the democratic style. More of the major projects are formulated and implemented in close partnership between the lord mayor and the central government before the public is informed. The abundant use of development corporations to implement large-scale projects reflects this change in policy style.

11.5 Conclusions

Copenhagen suffered from economic decline, social problems and worsening financial situation from the early 1970s onwards. The difficulties were further enforced by the political split, which for decades stopped all major initiatives. Real changes toward more proactive policies became possible through a coincidence of European and national policy, as well as changing local circumstances and the formation of a new governance style in Town Hall. The rising territorial competition among European metropoles in the wake of the decision for a Single Market and a slow withdrawal from the principle of equal regional opportunities gave national government a new understanding of the capital region as a key asset and a cornerstone in a geo-political strategy to place the country as a focal point for the emerging Baltic region, the Nordic countries and the Øresund region. The City itself changed from a reactive traditionalist policy style to a proactive entrepreneurialist style around 1990, just as the national parliament decided to boost the city by major infrastructure investments. The fixed link to Sweden was the centerpiece, but the City's city-annex, Ørestaden, was a central locality to house the benefits of Europeanization.

The experiences from the period after 1970 clearly showed that a proactive authority with responsibility and ability to implement its visions on behalf of the wider region was not to come unless the government system changed dramatically. Several important political parties prevented this. Instead, the only possibility was to seek closer cooperation between the City and the national government. This partnership has produced a series of important results during the last 15 years and has managed to fuel the City's economy. In reality, the partnership has given back the City its dominant position, despite the fact that only 30% of the metropolitan population lives inside its borders.

Planning has become a hot topic, but not as it used to be – primarily a technocratic discipline. The linkage of strategic planning and local economic development has given planning a new and central position in local politics. Financial sustainability is a prerequisite for strategic planning, but it is the implementation of the planning policy which should provide the benefits. The City has made the balance between finances and economy a cornerstone in its political strategy. Without a financially sustainable approach, political independence will be illusory.

Yet, urban politics has always been balancing on the edge between economic growth and social welfare. This is also reflected in present policies of the City. On the one hand, the City Council has clearly prioritized economically sustainable solutions, but on the other hand introduced a series of urban welfare initiatives which attempt to bring benefits to the least privileged groups. In the long run, this neighborhood improvement scheme will influence the social composition and thus create a social upgrading of the population.

References

Andersen, H.T. and Jørgensen, J. (1995) "City Profile: Copenhagen", *Cities*, 12(1)

Andersen, H.T., Hansen, F. and Jørgensen, J. (2003) "The Rise and Fall of Metropolitan Government in Copenhagen", *Geojournal*, 58

Boliger for alle (2001) *Boligpolitisk strategiplan for Københavns kommune 2001-04*

By ved vand *Report by the City of Copenhagen*

Engsplankontoret (1947) *Skitse til Egnsplan for Storkøbenhav*
Egnsplanrådet (1950) *Københavnegnens planlægning*
Forslag til København kommuneplan (2005) *Kommuneplanstrategi.*
Desfor, G. and Jørgensen, J. (2004) "Flexible Urban Governance. The Case of Copenhagen's Recent Waterfront Development", *European Planning Studies*, 12(4)
Harding, A. (1997) "Urban Regions in a Europe of the Cities", *European Urban and Regional Studies*, 4(4)
Harvey, D. (1989) "From managerialism to entrepreneurialism: the transformation of urban governance in late capitalism", *Geografiska Annaler B*, 70
Hovedstadsreformudvalgets sekretariat (1970) *Hovedstadsområdet. Befolkning. Opgave*
National Planning Report (1992) *Denmark towards the year 2018*

12. Regional Governance and Urban Regeneration: The Case of the Stuttgart Region, Germany

Johann Jessen

12.1 Outline of the Stuttgart Region

The Stuttgart region is placed in the heart of Baden-Württemberg, one of the 16 states of the Federal Republic of Germany. It comprises the city of Stuttgart, the state capital, and five counties (kreise) --- Ludwigsburg, Esslingen, Göppingen, Rems-Murr and Böblingen --- and covers 3,654 square kilometers. It is home to about 2.7 million people who live in a total of 179 municipalities; some of them are big towns, others small rural villages. With more than 700 inhabitants/square kilometer it is one of the most densely populated urban regions in Germany. Over 20 million people live within a 200-kilometer radius of Stuttgart.

The region is characterized by its distinctive polycentric urban structure. The relatively small core city of Stuttgart (590,000 inhabitants) is surrounded by several middle-sized towns that have very distinctive historic and economic profiles: Esslingen (89,000 inhabitants), Ludwigsburg (86,000 inhabitants), Sindelfingen (60,000 inhabitants) and others. It is located in the southwest Germany, within short traveling distance to France and Switzerland.

Compared to other German metropolitan regions, the Stuttgart region boasts a high income level and purchasing power, a relatively low unemployment rate (6.2% in 2006) and a booming real estate market. There is still a growing demand for housing and infrastructure. The labor market is attracting migrants nationally and internationally. Today, about 20% of the people living in the Stuttgart region have a foreign background --- the majority coming from southeast Europe and Turkey. The social and economic integration of foreign migrants will remain a major challenge.

Fig. 12-1. Stuttgart region in Germany (Source: Verband Region Stuttgart)

At present, the Stuttgart region is one of the economically most dynamic urban regions in Europe. Together with the Munich region it is part of the prospering South Germany. This economic strength is largely based on:

- Technology-led industrial clusters: Compared to other German regions, Stuttgart still has a strong industrial base. Stuttgart was the cradle of car manufacturing, and the automotive industry is still of great importance to the Stuttgart conurbation. This sector accounts for every sixth job in the region --- some 170,000 employees in total. A dense network of companies has been established over the years, comprising global players like DaimlerChrysler and Porsche, world-wide operating systems and component suppliers like Bosch, service providers, many research and development facilities, and consultant engineers. The industrial base has diversified in recent decades. The region is also home to a growing number of enterprises related to information and communication technologies like the German plants of SEL Alcatel, IBM and Hewlett Packard, environmental technology and the media. The specific mix of global players and innovative small and middle-size firms that successfully compete on the world market is one of the key factors behind the region's economic success.
- Outstanding research and training capacities: With two well reputed universities and a considerable number of private and public research

institutes, the region is one of the most productive nuclei of technological innovation, showing the largest number of patent-applications of all major urban regions in Germany. Moreover, the universities and a variety of other higher education institutions offer courses leading to both German and international professional degrees. This is one of the reasons for the high level of professional qualifications and skills characterizing the labor force in the Stuttgart region.

- Good accessibility: The region is well integrated in the dense German network of motorways, and is connected to the high-speed ICE rail network of the Deutsche Bahn AG (German railway company). Stuttgart Airport is the sixth largest in Germany. Together with its central location this allows for direct and comfortable access to markets and institutions throughout Europe.

Though still going strong, the region is aware of its vulnerability due to the relatively high dependency on the automotive industry. In the early nineties, the region experienced the first severe economic crisis after World War II. More than 56,000 jobs in the car industry disappeared. But the modernization and rationalization strategies taken up in the mid-nineties proved successful. Nevertheless, the structural change (deindustrialization process) will go on and will affect the Stuttgart region in future. One of the major objectives of regional policy is to diversify the economic base and make it less vulnerable mainly by strengthening the high profile of research and development and by keeping the region attractive not only for investment but also as a place to live and work. In this context, urban regeneration will become one of the main fields of urban policy in this region.

12.2 Regional Institutions: Verband Region Stuttgart

The history of the Stuttgart region as an administrative entity dates back to 1974, when the so-called Regionalverband Mittlerer Neckar (Mid-Neckar Regional Association) was established. It was basically a planning institution responsible for setting up the formal Regional Plan. In the early nineties, the political debate on strenghtening regional-level political institutions in the Stuttgart conurbation started off with a number of regional conferences involving all relevant actors. This was part of the answer to the structural crisis emerging at that time.

As a result, the Verband Region Stuttgart (VRS) was founded in 1994, based on a state law that highlighted the regional planning association for the capital region as a special entity by providing it with more political power than the other regional planning associations in the state of Baden-Württemberg. This extended power of the VRS is based on the direct election of a regional parliament (Regionalversammlung) with proportional representation of the central city and the five counties. The Regionalversammlung was the first directly elected regional parliament in Germany (5-years term), and since 1994 only the Hanover region, *Großraum Hannover*, has been added. The decisions of the regional assembly are carried out by the VRS regional office (with a total of 45 employees).

The VRS has the following tasks and responsibilities (www.region-stuttgart.org): Regional planning, transport, environment, economy.

- Regional planning: The VRS sets up the Regionalplan (Verband Region Stuttgart 1998), which is a formal regional development plan that formulates the goals, basic principles, and suggestions from which the planners at the municipal level have to take their line. The plan defines zones of housing and commercial development, as well as green belts and zones that are to be kept free from development. The present Regionalplan (with a time-horizon of some 10 to 15 years) was adopted in 1998, and its updating is underway.

Fig. 12-2. Verband Region Stuttgart: Administrative units (Source: Verband Region Stuttgart (1998))

- Transportation: The VRS region has developed an integrated traffic and transport concept which is coordinated with the regional plan and landscape plans. Finalized and approved in 2001, it contains long-term strategies for roads, railways, and bicycle paths, and a list of priorities for future investments. The VRS is the authority for the planning and operation of regional public transportation (*S-Bahn*) and thus the organising body behind the local suburban electric railway system as well as, since October 1999, all regionally important public transportation.
- Environment: The VRS is responsible for landscape planning as an important component of regional planning. It has taken up a more proactive role in landscape planning by drawing up a general concept called the Greater Stuttgart Landscape Park (Verband Region Stuttgart 1994), showing where green open spaces are to be linked up together and to be improved. The basic idea of the regional landscape park is to enable people to enjoy the landscape, and protect it at the same time. Detailed plans of smaller intermunicipal landscape parks (Landschaftspark Neckartal/landscape park Neckar Valley, www.landschaftspark-neckar.de and Landschaftspark Filder/landscape park Filder) are under way. The upcoming implementations are organized as joint ventures of the region, the municipalities and various other authorities.
- Economy: Business development activities concentrate on establishing industry clusters and extending existing ones, as well as building up networks to connect companies and research establishments within the Stuttgart region. The VRS has established an independent agency for economic promotion (Wirtschaftsförderung Region Stuttgart/WRS, www.wrs.region-stuttgart.de, Stuttgart Region Economic Development Corporation), that closely cooperates with the chamber of commerce. It provides support to companies that want to settle down in the region as well as to local companies wishing to expand by helping them find suitable commercial and industrial sites and real estate. Moreover, the VRS has initiated another independently operating agency focusing on tourism marketing (Regio Stuttgart Tourismus Marketing GmbH, www.stuttgart-tourist.de).

Furthermore, the VRS has the right to take over responsibility for tasks in the fields of sports, culture, congresses and trade fairs at the scale of the urban region, and it makes use of this authority in various ways. There are, for instance, joint activities with the municipalities such as the KulturRegion Stuttgart or the SportRegion to promote cultural life and sport events on a regional level. And there are private-public intiatives such as the Forum Region Stuttgart (together with the Chamber of Industry and Commerce) that promote the region as an economic, cultural and social asset. Another example may be the Dialogforum der Kirchen, which serves to improve the dialogue between the Churches of the region.

The VRS acts as a shareholder in various agencies and companies: 51% of the WRS, 32% of the tourism marketing company, 20% of the VVS (Public Transport Network) and 10% of the New International Trade Fair Company.

The annual budget of the *Verband Region Stuttgart* amounts to €263 million (2005). The VRS is financed approximately half-and-half by grants from federal and state government (35%, 15%) and transfers charged from the member counties and municipalities (40%, 10%) in the region. By far the major item of expenditure is on local public transport, which takes up about 85% of the budget.

There can be no doubt (and it is confirmed by regional research studies: Benz (2003), Knieling (2003), Kübler (2003)): the Verband Region Stuttgart is an outstanding example - both on a national and international level - of a regional actor that combines traditional formal instruments of regional planning and institutionalized policies with new strategic elements that are usually associated with regional governance: strategic planning, forming networks, promoting cooperative planning, creating new forms of private-public partnerships, and so on. Thus, it can be said that the Verband Region Stuttgart is one of the spearheads of new and innovative concepts of regional governance in Germany.

12.3 German Planning System and Sustainability

Germany is a federal state with three levels of government: the federal government (*bund*), the states (*länder*), and the municipalities (*gemeinden*) (Bundesamt für Bauwesen und Raumordnung (2000)). The Constitution, written after World War II, clearly defines the division of power among the three levels. The planning system in Germany is characterized by a "strong legal framework and a decentralized decision-making structure" (Newman/Thornley (1996)). On the central state level, there are only general legal guidelines. The legal framework consists of the Federal Regional Planning Code (*Raumordnungsgesetz*) and the Federal Building Code (*Baugesetzbuch*) which provide the legal instruments for regional planning on the state level and for urban planning on the local level. This framework ensures a certain degree of consistency of planning on the lower levels (*länder* and *gemeinden*). Each German state organizes regional

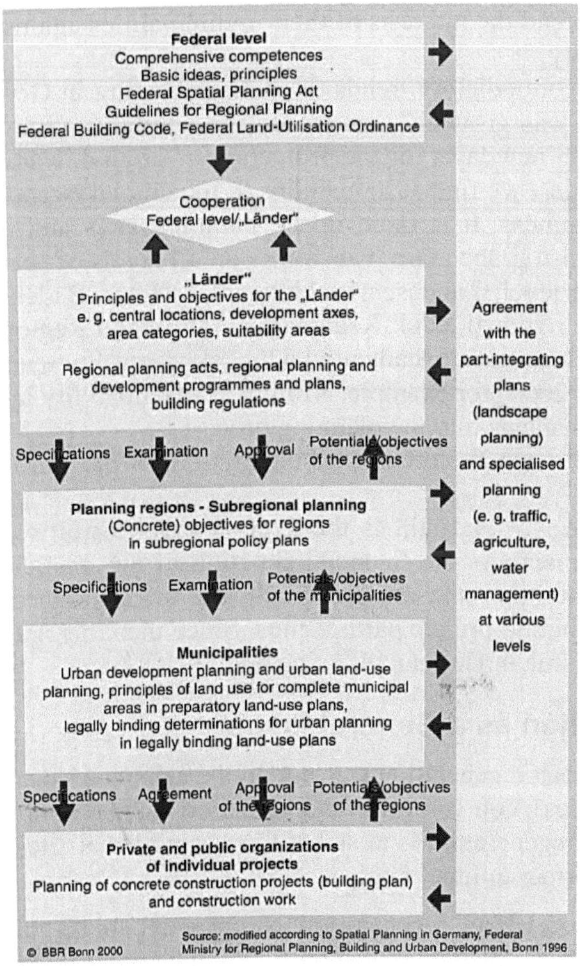

Fig. 12-3. Planning levels in Germany (Source: Bundesamt für Bauwesen und Raumordnung (2000))

planning differently and, as the example of Baden-Württemberg shows, even within one state it may differ. The Verband Region Stuttgart represents a stronger planning body than the other 12 regional planning agencies in Baden-Württemberg, which are mostly restricted to formal regional planning.

Traditionally, German municipal governments enjoy a high degree of autonomy in planning and have strongly defended this against any demands for centralization of power on a regional or state level. The influence of the counties (*kreise*) on spatial development is significant at the regional scale since they defend the interests of their municipalities against the region and the central city.

The last decade showed contradictory trends of spatial planning in Germany. On one hand, there was growing concern for the natural environment, which was reflected in new laws for the protection of ground, water and air. Since the mid nineties, we find sustainability as the crucial overall objective of spatial development. It is fixed in the planning laws and it forms the overarching premise in the main plan works on all levels of spatial planning - on the state level: Landesentwicklungsplan (Land Baden-Württemberg 2002); on the regional level: Regionalplan (Verband Region Stuttgart 1998) and on the local level: Stadtentwicklungskonzept Stuttgart (Landeshauptstadt Stuttgart 2004) for example. Aiming at sustainability introduced more regulatory elements into the planning process.

This trend is enforced through the process of integration on the European level.

On the other hand, there was and still is the general trend towards a more property-led development. As the financial situation of the municipalities (*gemeinden*) became more and more difficult, big urban projects and events were set out as public-private partnerships. Since then, deregulation has become a major issue in German planning policy.

12.4 Urban Regeneration as a General Issue

In Germany, the issue of urban regeneration was put on the agenda of local planning in the early eighties. You can differentiate between three major types of sites where urban regeneration is at stake: former industrial sites, former railway sites, and former military sites.

1. Former industrial sites: Urban regenaration is supposed to be the urban policy and local planning response to the process of deindustrialization in the Ruhr and Saar areas, major urban growth poles in the fifties. These old industrial regions faced severe economic decline of their basic industries (steel production and mining), even though extensive subsidies were spent to cushion the process. Other old industrial regions followed: harbor cities in the north that faced the breakdown of the shipyard industries, or former centers of the textile industry. For the first time, urban decay and stagnation with extensive areas of derelict industrial land were (and still are) a major urban planning issue in Western Germany. The International Building Exposition Emscher Park in the Ruhr area, which started in 1988 and ended in 1999, served as a hothouse to develop comprehensive regional strategies and to define and test new concepts to modernize old industrial regions.

2. Former military sites: After the German reunification and the withdrawal of the Soviet, American and Allied Forces, the restructuring or ′conversion′ of military areas offered enormous potential for urban regeneration. Spacious areas, once shut off from the public even for decades and centuries, are now ready to be integrated into the existing urban pattern. Almost every German town had at least one former military area that awaited new uses. This is also the case in the Stuttgart region, and some of the garrison towns in former East Germany even have an abundance of abandoned military sites that are at disposition.
3. Railway sites: Great changes in international and interregional accessibility of urban regions as well as within the urban pattern were affected both by the new high-speed train system ICE (Intercity Express), which was introduced in the late eighties and will be extended in the forthcoming years, and by the privatization of the Deutsche Bundesbahn (German federal railway), now Bahn AG, in the early nineties. The restructuring of the railway system and railway station areas gave way to ambitious plans for urban regeneration around the old railway sites.

In Germany, urban regeneration projects are seen as local projects, not as regional projects, though some of them may have tremendous regional impacts. Consequently, the most important actors in this field are the municipalities.

As urban regeneration projects are very complex and demand a high amount of intermediate inputs, as well as organizational and management capacities, there is a broad consensus that this exceeds the resources of most of the municipalities. That is why the federal government and various states provided funding programs (Förderprogramme Stadtumbau Ost und West) for urban regeneration. As the problems accumulate in eastern Germany, most of the money goes into urban regeneration projects in eastern German cities. There are also special legal planning and financing instruments at hand that can facilitate the implementation of regeneration projects (Stadtentwicklungsmaßnahme; Städtebaulicher Vertrag).

12.5 Urban Regeneration in the Stuttgart Region

At present, we have a great number of urban regeneration projects in the Stuttgart region, mostly on former industrial and military sites. Almost all local authorities in the region of Stuttgart follow the guideline of "brown field development before greenfield development." Thus, most of the urban regeneration projects are important elements of the local development strategies towards sustainability. Urban regeneration is a major focus of Stuttgart city's new masterplan (LandeshauptstadtStuttgart (2004)).

The urban regeneration projects are initiated and managed by those municipalities that have the site within their administrative boundaries. Due to the booming real estate market, most of the former industrial sites can rather easily be converted to new uses, be it for commercial, public facilities or new industries, even when an expensive cleanup of toxic waste is necessary. Many of those projects get funds and other kinds of support from the state (Wirtschaftsministerium Baden-Württemberg (2000)). The regeneration of the Roser Areal in Stuttgart-Feuerbach is a good example. On the site of a former textile and leather factory, a new urban quarter with a mix of commerce, housing and services has been created. This is an example of a more or less market-driven urban regeneration project.

There are also many urban regeneration projects on former military sites. The most outstanding one is the conversion of former American barracks in Ostfildern city into the biggest newly built housing area in the state of Baden-Württemberg: the Scharnhauser Park --- a housing area for more than 12,000 people. Ostfildern is a town with about 32,000 inhabitants in the south of the region at the border of Stuttgart.

Regional institutions such as the VRS are not directly involved in these urban regeneration projects, even when they have significant regional impact. But the projects comply with the regional planning policy. On the regional level, urban regeneration is predominantly supported indirectly by a formal planning framework and by information and coordination:

- Regional planning: The present official Regionalplan (Verband Region Stuttgart (1998)) has urban sustainability as an overall objective. The landuse planning concept gives high priority to brownfield development.
- Landscape planning: Protection of the natural resources and landscape is a crucial element of regional landscape planning.
- Information and coordination: Providing an information system on the potential for brownfield developments in the region (Verband Region Stuttgart (2005)).

12.6 Stuttgart 21

The urban regeneration project with the greatest impact on regional development is without any doubt the project Stuttgart 21 (Turmforum Stuttgart 21 (2005)); (Wolfrum (2003)). Stuttgart 21 is the marketing name for the regeneration of the railway site at the main station of Stuttgart. It is the only urban regeneration project the Verband Region Stuttgart as a regional institution is engaged in. The Neue Landesmesse (New Trade Fair) at the Stuttgart Airport, currently under construction (opening fall 2007), is another very important project with tremendous regional effects, in which the VRS is directly involved as one of the shareholders of the new international trade fair company. But the Neue Landesmesse can hardly be considered a contribution to urban regeneration, as it is a pure greenfield development. It is built on the most fertile agricultural land the region has to offer. This fact was subject to a long and fierce argument between the institutions (state, VRS and the city of Stuttgart) that promote and finance the project on one hand and environmentalists, local farmers and the affected municipality of Leinfelden-Echterdingen on the other hand.

In April 1994, the chief executive of Bahn AG (German railway company), the mayor of Stuttgart and the respective transport ministers of the region and the federal government presented the proposal for Stuttgart 21

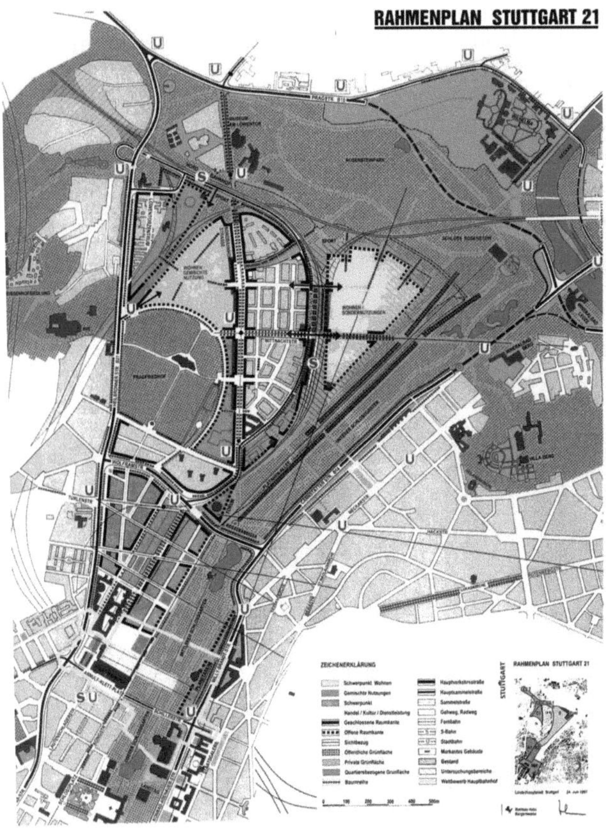

Fig. 12-4. Stuttgart 21 master plan (Source: Landeshauptstadt Stuttgart (2004))

to the public. The proposal included three basic elements:

1. new railway tracks and nodes for long-distance and local public transport including a second new terminus for the high-speed train at the airport and the new trade fair;
2. a new urban quarter (100 ha) to be developped on the railway land for 10,000 people and 24,000 jobs;
3. a new underground through-station replacing the existing cul de sac main station.

This concept was developed and refined in the mid-nineties. The results of various elaborate competitions and participation processes were fixed in a master plan that is still valid and represents the basis for all the detailed technical planning and legal planning procedures that have already been carried out or are still to come. Though the vast majority of the local parliament supports Stuttgart 21, the project is still a matter of great controversy; environmentalists and the Green Party in particular are still opposing the project.

Stuttgart 21 is the most ambitious endeavor of the region. There are high expectations. Project Stuttgart 21 will:

- integrate the Stuttgart region into the European high-speed train network (Paris-Munich-Vienna) and substantially reduce long distance travel times;
- contribute to an environmentally sound transport system;
- increase the accessibility and the attractivness of the airport and trade fair;

Fig. 12-5. Model of Stuttgart 21 (Source: Landeshauptstadt Stuttgart (2004))

- give way to more sustainable regional development by developing a new inner-city area of 100 hectares, by expanding inner city parks (Schlossgarten and Rosensteinpark) and by substantially improving the public transport system;
- make the city of Stuttgart a more livable place by providing new attractive locations for housing, services, retail and cultural venues (including architectural flagships such as the new underground railway station and the so-called Library 21) and by linking up those parts of the inner city that have been separated by the railway tracks for decades.

The costs of the project will amount to €2.8 billion. The Bahn AG, the state of Baden-Württemberg, the federal government and the VRS signed an agreement that fixed the obligations of the project partners. The Bahn AG will invest €1.415 billion. The company expects to cover this funding by the revenues from the sale of land, which today seems to be an optimistic, if not risky assumption. The remaining €1.785 billion will be financed

Fig. 12-6. Stuttgart 21: Bird's eye view (Summer 2005)

through subsidies and loans from the federal government (€0.5 billion, including € 0.365 billion from the European Union), the state of Baden-Württemberg (€0.635 billion), Stuttgart Airport (€0.15 billion) and VRS Region Stuttgart (€0.1 billion). The financial risks of Stuttgart 21 of about €1.32 billion are borne by Bahn AG (€3.80 billion), the state of Baden-Württemberg (€0.532 billion), the city of Stuttgart (€0.279 billion) and Stuttgart Airport (€0.129 billion). Closely connected with the Stuttgart 21 project is the construction a new railway track for the high-speed train between Stuttgart and Ulm (part of the high-speed line between Paris and Vienna) which will cost about €2 billion that is to be shared between the federal government (€1.05 billion, including up to €0.1 billion from the European Union) and the state of Baden-Württemberg (€0.95 billion).

The VRS considers it as an important goal to improve international and regional accessibility. Since the VRS is responsible for regional public transport, it is supporting Stuttgart 21 with a total contribution of €100 million. Moreover, the VRS has incorporated the projects Stuttgart 21 and the new trade fair into the urban regional development plan and the transport structure plan and thereby provided the legal planning framework. However, all in all, the contributation of the region to the project Stuttgart 21 should not be overestimated. The most important actors involved are the Bahn AG and the city of Stuttgart.

The implementation of Stuttgart 21 is notably behind schedule. Up to now, only a few buildings adjacent to the railway station have been completed. Construction work on the tunnels should have started in 2005. After a very long and complicated bargaining process, the final decision on the financing and the share of costs and risks between the shareholders was made in July 2007, when the final agreement was signed. Construction will begin in 2010. The opening of the new railway station will be in 2019.

12.7 Sustainability, Regional Governance, and Urban Regeneration: Conclusions

My conclusions are based not only on the case study of Stuttgart presented here; they reflect also findings of a comparative study that was carried out on recent urban policies in eight European regional cities, i.e. metropolitan regions at the European scale, among them Barcelona (Spain), Manchester (Great Britain), Amsterdam (Netherlands), Zurich (Switzerland) and Leipzig (Germany) (Jessen et al. (2007)). The conclusions are summarized in six points.

Urban Regeneration as a Major Policy Issue

In regional cities all over Europe, urban regeneration has been a major political issue on a local level for about two decades. This holds true both for ever dynamic and prospering cities like Barcelona, Zurich, Copenhagen, Munich, or Stuttgart and for cities in (formerly) declining regions like Manchester, Glasgow, Lille or Leipzig in former East Germany. In Germany, urban regeneration was put on the agenda of local planning in the early eighties. In the wake of globalization, very different processes opened up the window of opportunity for inner-city regeneration: (1) deindustrialization, (2) transformation of the welfare state (deregulation and privatization) and (3) finally - very important for Germany - the end of the cold war and the process of European integration through generating new options for urban development on highly accessible, abundant land such as former industrial sites, harbor sites, railway sites and military sites.

Sustainability as an Overall Objective

The overall objective of sustainability has been attached to urban regeneration since the early nineties in the course of the debate on the outcome of the UN Conference in Rio de Janeiro. Urban regeneration has to find a balance between economic, social, ecological and cultural objectives. This requires an integrated approach. Urban development strategies with different goals such as increasing regional economic competitiveness, preserving a city's quality of life, protecting the environment and conserving energy have increasingly become interconnected. European cities that are renowned for their urban policy follow an integrated perspective to enhance their economic, cultural and ecological standing. This is partially rooted in the global decline of manufacturing and the rise of knowledge-based urban economies. These cities share the ability to attract human capital necessary for knowledge-based industry sectors such as high-technology, entertainment and financial services. They adopt a two-pronged approach to promote economic competitiveness and quality of life at the same time, highlighting the connection between both. In practice, this often leads to a dilemma: economically successful regions become growing regions and as a result have to deal with intensifying land uses, urban sprawl, and greater population densities and diversity. At the same time, the locally successful knowledge-based economies attract well-educated employees who in turn demand quality of life for themselves and their families. The reasons for the increased importance of quality of life factors in a post-industrial society stem from the fact that location decisions of firms and households no longer depend on the physical endowments of a place that used to be important for the manufacturing sector (like the access to natural resources). Urban characteristics such as ethnic diversity, an innovative and entrepreneurial environment, as well as a region's ability to curb urban sprawl and preserve open space, or opportunities for high-quality affordable housing play a more important role today. Studies have shown that the so-called creative class (Florida, 2002) within the knowledge economy are attracted to regions that offer the right mix of these factors. Urban regeneration often is the main strategic element of urban policy to promote this mix.

Strategic Concepts and Urban Regeneration

The European experience shows that for the success of urban regeneration, overarching urban development concepts (guidelines, mission statements) are indispensable. In order to actively plan for the future, cities need an overarching vision of urban development to which strategies and policies can be aligned. In the competition for resources – may it be companies, residents, purchasing power, subsidies or public attention – guiding models are very important as an instrument for profiling a city, as well as for communicating the objectives to the public. The cities we studied clearly underline this dual function of an urban vision, although their plans for the future vary in implementation, in the extent to which they are binding, and in the role they play in the political decision-making process, marketing strategies and image campaigns.

The main objectives of almost all strategies for urban regeneration projects are promoting mixed usage and re-urbanization. Often, housing projects have a leading function and are offered in various forms. Whether through demolition, new development or conversion of historical buildings, offering a housing supply that is tailored to the needs of specific target groups is a key to success. Our case studies have shown that new inner-city housing forms are comprised of exclusive lofts in old brick buildings and luxury apartments in high–rises in the higher price range as well as small living compartments in blocks of flats; and recently accommodation for families with children. Space is often also offered for firms in the media, entertainment and design fields. In many cases, temporary uses are no longer regarded as a disturbance or handicap, but rather embraced as cultural pioneers that are essential for reintegrating previously "forbidden or forgotten" areas. They create new images of the area that prepare the ground for actual restructuring and facilitate the search for builders and future inhabitants.

Urban Regeneration and Cooperative Planning

Urban policies in general and urban regeneration strategies in particular are characterized by locally defined constellations consisting of very different stakeholders: city parliament, city administration, companies, project developers, foundations, citizens, alliances, associations, private individuals, and last but not least, higher-level institutions such as regional and central government bodies and the European Union. Over the past years, the division of labour and the balance of power between the public and private sector has largely changed. New forms of organizations (partnerships, alliances, clearing houses, etc.) have emerged. They pool information, carry out the actual coordination processes, make arrangements for important decisions, and decide on operative implementation steps. This has also redefined the balance of power between the municipal administration and the local parliament. Close collaboration between elected officials and the highest level of local administration is required, as no urban development can succeed without political backing. Traditional forms of citizen's participation and local parliaments seem to lose ground.

Urban Regeneration as Part of Multi-level Policies

As urban regeneration projects are very complex and demand high financial resources and management capacities, there is a broad consensus in many European countries that this exceeds the budget of most municipalities. This is why the governments provide funding programs for urban regeneration, and in the case of Germany there are two programs, one for eastern Germany and one for western Germany (Förderprogramme Stadtumbau Ost und West). As problems accumulate in eastern Germany, most of the funding is spent on urban regeneration projects in the cities of the east. There are also special legal planning and financing instruments at hand that can facilitate the implementation of regeneration projects. The municipalities can get support through funding programs at both state and federal levels. Specific legal instruments for land mobilizing are also provided. Recently, subsidies from various European Structural Funds are becoming more and more important. Many of the urban regeneration projects would not have gotten off the ground without support from the European level.

Urban Regeneration and Regional Planning

In Europe, urban regeneration projects are seen as local projects, not as regional projects, though most of them may have tremendous regional impacts. Consequently, the most important protagonists in this field are the municipalities. While in many cases the organization on the local administrational level has been thoroughly modernized and remodeled, within most of the large European regional cities regional planning and cooperation still have a minor influence. The picture is still dominated by particularism and rivalry between the core city and neighbouring communities. Existing regional cooperation is often poorly formalized and the few legal-based regional institutions that exist have even been weakened in many places in recent years. The case of Stuttgart with its relatively strong regional planning association and its ambitious regional policy is an exception both on a national level and European level. The Verband Region Stuttgart (VRS) has a very broad understanding of regional policy. Since its establishment in 1994, it represents one of the most advanced forms of supra-municipal planning and regional governance within the Federal Republic of Germany. The aim of developing the region on a sustainable basis is one of the guiding principles for all of its activities. The VRS plays an active role in many fields of regional policy, if not so much in the field of urban regeneration. The formal regional planning and the regional landscape policy of the VRS follows the guideline of "brownfield before greenfield development," and therefore co-stimulates urban regeneration projects on a local level.

References

Benz, A. (2003) "Regional Governance mit organisatorischem Kern. Das Beispiel der Region Stuttgart", *Informationen zur Raumentwicklung*, 8/9

Bundesamt für Bauwesen und Raumordnung (2000) *Urban Development and Urban Policy. An Overview* (Berichte No. 6)

Jessen, J., Meyer, U. and Schneider, J. (2007) stadtmachen.eu. Stuttgart: Zurich (publication forthcoming)

Knieling, J. (2003) "Kooperative Regionalplanung und Regional Governance: Praxisbeispiele, Theoriebezüge und Perspektiven", *Informationen zur Raumentwicklung*, 8(9)

Kübler, D. (2003) "'Metropolitan Governannce' oder: Die unendliche Geschichte der institutionennbildung in Stadtregionen", *Informationen zur Raumentwicklung*, 8(9)

Landeshauptstadt Stuttgart (2004) *Stadtentwicklungskonzept. Entwurf 2004*

Land Baden-Württemberg (2002) *Landesentwicklungsplan Baden-Württemberg 2002*

Newman, P. and Thornley, A. (1996) *Urban Planning in Europe*. International Systems and Planning Projects

Turmforum Stuttgart 21 (ed) (2005) *Das Projekt Stuttgart 21*

Verband Region Stuttgart (1998) *Regionalplan 1998*

Verband Region Stuttgart (ed) (2005) *Nachhaltiges regionales Siedlungsflächenmanagement in der Region Stuttgart. Schlussbericht des Forschungsprojekts* MORO-RESIM
Wirtschaftsministerium Baden Württemberg (ed) (2000) *Recycling von Brachflächen in Baden-Württemberg*. Informationen für Städte und Gemeinden
Wolfram, M. (2003) *Planning the Integration of the High-Speed Train: A discourse analytical study in four European regions* (Dissertation at the Faculty of Architecture and Urban Planning of the University of Stuttgart)

Landeshauptstadt Stuttgart, http://www.stuttgart.de (Accessed in 2007)
Landesmesse Baden-Württemberg, http://www.landesmesse.de (Accessed in 2007)
Landschaftspark Neckar, http://www.landschaftspark-neckar.de (Accessed in 2007)
Stadt Ostfildern, http://www.ostfildern.de (Accessed in 2007)
Stuttgart 21, http://www.stuttgart21.de (Accessed in 2007)
Verband Region Stuttgart, http://www.region-stuttgart.org (Accessed in 2007)
Wirtschaftsförderung Region Stuttgart, http://www.wrs.region-stuttgart.de (Accessed in 2007)

Acknowledgements

I would like to thank Karoline Brombach (Institute for Urban Planning, University of Stuttgart) and Dr. Jürgen Ludwig (Verband Region Stuttgart) for their helpful comments on the draft.

13. City-Region Planning and Urban Projects "Coexistence or Integration?": The Case of Marseille-Aix in France (1990-2006)

Alain Motte and Jérôme Dubois

13.1 Overview of the Marseille-Aix City Region[1]

13.1.1 Organization of the City Region and Main Issues

The Marseille-Aix city region developed in a geographic location which conditioned its urban organization. This Mediterranean littoral territory is structured by coastal mountains (Cassis la Ciotat, Calanques, Côte Bleue), fragile humid spaces (Etang de Berre, Camargue), and a river (Rhône and its distributory La Durance). The land littoral itself has constraints: the mountain amphitheater of the Marseille *commune*, Alpilles and Lubéron. All these elements have imposed the circulation corridors and influenced the urbanization processes.

A Marseille metropolitan planning scheme was elaborated in the late sixties and the early seventies by local representatives of the central state in consultation with *communes*. This scheme was used by the central state to coordinate its major investments in the urban region. It has also had a major influence on the urbanization processes in this urban area since the seventies, mainly through the highway system.

No metropolitan strategy has been elaborated since the seventies. The decentralization process during the 80s and 90s spurred competition among communes. The local institutional culture is historically one of autonomy and non-cooperation[2] between *communes*. Since 1990 the central state, using institutional national reforms (1992 and 1999-2000) and specific policies, has worked toward the emergence of metropolitan policies in the Marseille-Aix city region.

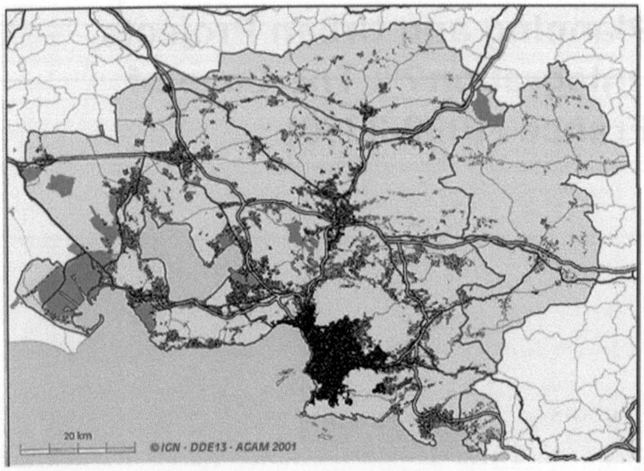

Fig. 13-1. Urbanization and highway structure of the Marseille-Aix city region (2000)

What should be acknowledged first is the fundamental role of the central state and its representatives in the city region. This is not a role of the imposition of policies. Since the beginning of the decentralization reforms of the 80s the central state has been encouraging local public institutions (*commune*, *département*, *région*) to voluntarily adopt (Thoenig, Duran, 1996) new policies.

How has it been possible for the territorial public institutions to increasingly cooperate during the 90s? What has been the precise role of the central state and its representatives? Has this role been carried out specifically through national reforms? Have any city-region policies developed?

The Marseille-Aix city region is organized around the territories of the main *communes* in the area. The *communes* of Marseille[3] (797,491 inhabitants in 1999) and Aix-en-Provence (134,324 inhabitants in 1999) are the main centers of the city region. They had very different dynamics during the 80-90s period. Marseille began a severe decline and Aix showed economic growth[4].

The Marseille-Aix city region has changed rapidly during the 80-90s period due to many factors, the most important as follows:

- changes in productive systems linked to globalization developments
- residential restructuring (peripheral developments, village growth)
- social and spatial segregation developments.

In the absence of a metropolitan strategy during the 80s and 90s the metropole structured itself as a muli-polarized space, mainly on the basis of the 1970 Planning Scheme and the highway system built during the 60s and 70s.

The Marseille *commune* has the largest population of the city region but doesn't dominate it. Rather, the central city and peripheral cities not only function independently, but the development dynamics are very different. Growth during the 80s and 90s took place outside Marseille. Marseille was declining demographically and economically. Urbanization happened outside the Marseille territory

Residential changes become very clear when one reads the demographic statistics of the Bouches du Rhône *département*, in which Marseille and Aix-en-Provence are located. The population of the *département* as a whole grew from 1.719 million inhabitants in 1982 to 1.835 million inhabitants in 1999[5], but the area of the département outside Marseille has grown 8.2% (increase of 188,911 inhabitants) whereas Marseilles has lost 6.7% of its population (decrease of 73,360 inhabitants).

Housing growth outside Marseille is segregative. The urbanized space is structured by fragmented residential spaces (individual/collective, subsidized/non-subsidized housing) in a liberal mode with competition between the *communes* to attract the most affluent population. Residential redistribution is directly linked to the employment poles, as families are reducing their daily work-home movements (67% of individuals are working in their residence *commune*, compared with 30% in France)[6]. The residential periurban growth is based on individual housing. The residential urban growth is based on collective housing.

From an economic point of view the Marseille-Aix city region is organized around the following three main production areas, each with its own development dynamics.

- Marseille area, which collapsed during the 60s and 70s and dragged the city into a severe decline during the 80s and 90s;
- Fos-Martigues area (west of Marseille) dominated by large traditional industrial plants (chemicals, fuel refineries, gas plants), which developed mainly during the 60s and 70s and were under severe economic threat during the 70s and 80s;
- Aix area (Les Milles, Rousset, Saint Paul les Durance) with high technology development based on central state investments during the 60s and 70s (CETE[7] in 1968; CEA[8] in Saint-Paul in 1969), and the engine of the urban development processes during the 90s.

Two other areas have to be taken into account, as their dynamics have been very important for the city region.

- Vitrolles and Plan de Campagne, near the airport and the TGV rail line, with big industrial and commercial zones developing since the 60s;

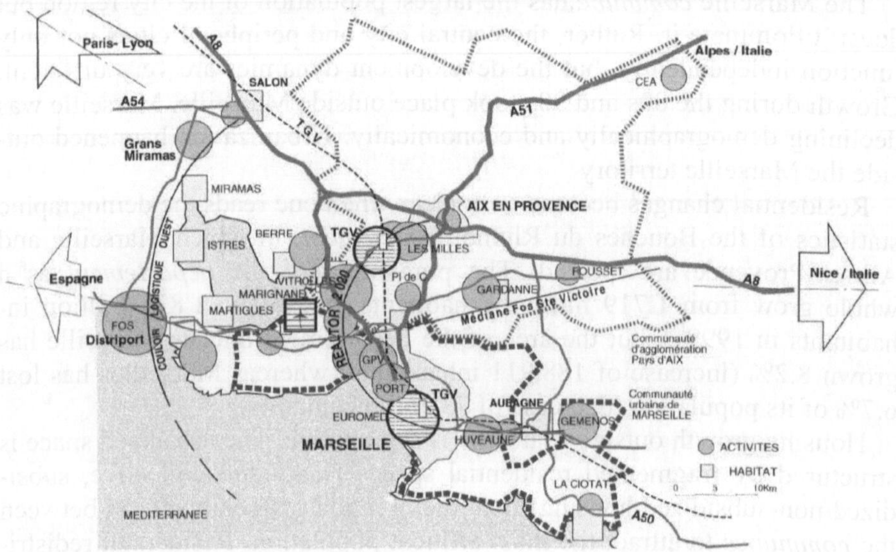

Fig. 13-2. Planning issues and governance in the Marseille-Aix city region (Source: Michel Chiappero)

- Aubagne and Gémenos (east of Marseilles), which received industries relocated from Marseille during the 80s.

Two other areas with traditional industries are under threat:

- Gardanne, a mining city between Aix and Marseille (mine closed in 2002);
- La Ciotat, east of Marseille, whose shipyards closed at the beginning of the 80s.

The structuring/re-structuring of productive centers is fundamental, and mostly explains the residential changes.

The city region has a concentration of national and international road, rail and air infrastructure. It is at the crossroads of a north-south axis which ends on the Mediterranean coast, and a west-east axis linking Spain to Italy.

The central state was strong during the 1960s and organized the Fos area development. It built the highway system which focused on road transport between the Rhône valley and Spain and Italy. This highway system has also contributed greatly to the urbanization processes of the city region since the 70s in facilitating residential developments around the main cities[9].

The metropolitan railway infrastructure has been largely abandoned, due to the priority given to the car between the 1960s and the end of the 1990s. Only the national level has taken rail into account, with the construction of a TGV line at the end of thc 90s.

The consequences of these policies are shown in a 1997 commuting survey: 91% of trips are made by private cars. The motorway system is saturated during peak hours every day, particularly the Marseille-Aix motorway, which is the spinal column of the network.

The Marseille-Provence airport is the third largest in France, with 6.4 million passengers in 2000[10]. Its development was very fast during the 80-90s period.

The Marseille-Aix city region also has many high-quality environmental and landscape spaces: Camargue, Alpilles, Sainte-Victoire mountain, Côte Bleue, Frioul isles, Calanques, Cap Canaille to name just a few. The most prestigious spaces are legally protected, but the general environmental quality of the city region declined during the 80-90s period due to the urbanization processes.

13.1.2 Evolution of the Local Government Administrative Structure

France is a unitary state, which built its local institutional system (*communes* and *départements*) during the Revolution (1789-1799) and the Consulat and Empire (1799-1814) based on highly centralized functions. Local authority was in the hands of the préfet, the local representative of the national government. Local representatives were progressively introduced and given authority during the 19th century, culminating in 1871 (*départements*) and 1884 (*communes*), the préfet still having a crucial role. Between the end of the 19th century and the decentralization measures begun in 1982, local authority was structured along two networks: the administrative network dominated by the préfet, and the political network based on decentralized local institutions (*communes, conseil général*[11]), the two networks working within centralization processes.

The dramatic evolution of post-World War II French society, and its rapid urbanization and industrialization, increased the political and administrative importance of mayors, particularly in cities (Bordeaux, Lille, Lyon, Marseille). The period from 1980 to 2000 saw dramatic change in the development of powers of local elected representatives, and the role of the *préfet*. The first main reform of 1982-1983 moved the center of gravity of French institutions from central spheres to local and regional spheres, and from the administrative network to the political network. This

Table 13-1. The level of governance in France (2006)

Level	Politics	Planning Responsibilities
Nation	Election	9 National Planning Schemes*1)
Région	Election	Schéma régional d'aménagement*1) Nation / Region (2000-2006) Plan Contract Nation / Region (2007-2013) Project Contract
Département	Election	
Communautés	Representatives of Communes	Région / Communautés contract Schéma directeur (1982-2002) or Schéma de cohérence territoriale*2) (from 2002), Plan local de l'Habitat (Housing Development Plan), Plan de déplacement urbain (Mobility Development Plan)
Communes	Election	Plan d'occupation des Sols (1982-2002) or Plan Local d'Urbanisme*2) (from 2002)

1) Loi « Voynet » : Loi n° 99-533 du 25 juin 1999 (J.O. du 29 juin 1999, pp. 9515-9527) and Loi « Pasqua » : Loi n° 95-115 du 4 février 1995 (J.O. 5 février 1995, pp. 1973 & suiv.)

2) Loi « Gayssot-Besson » : Loi n° 2000-1208 du 13 décembre 2000 (J.O 14 décembre 2000, pp. 19777 & suiv.)

decentralization processes maintained the same institutions, but changed their roles within the new system.

The local and regional authorities (*communes*, *departments* and *régions*) have largely developed their responsibilities and now have the capacity to define their own agendas. The *commune*'s role has been reinforced as manager of the local territory (planning permission, *Plan d'occupation des sols* or *Plan Local d'urbanisme*, *Schéma directeur* or *Schéma de Cohérence Territoriale*). The département is no longer directly dependent on the préfet. It has its own elected president (selected by *département* representatives), and has developed its importance in education (investments in secondary schools), roads and transportation, and social and sanitary policies. The *région* was transformed into a full Public Territorial Authority with a president elected by a regional council[12]. The *région* has a crucial role in coordinating investments of the local authorities (regional planning and *aménagement du territoire*), education (investments in secondary schools), professional training and transport. All these authorities can also promote economic development.

The préfet (national government representative) role

With reduced power over the local and regional authorities, the *préfet* (local and regional representative of the national government) has become an intermediary, encouraging local authorities to adopt national policies. The *préfet* is also in charge of controlling the legality of the public authorities' actions, and acts mainly to coordinate the actions of the administrative network at *région* and *département* levels. The central ministries are still working on a vertical basis and the horizontal functioning at regional and *département* levels are difficult to put into practice. This is weakening the position of the *préfet* because territorial problems have to be managed on a horizontal basis by the local authorities, and only the communes, *départements* and *régions* have the capacity to coordinate and establish hierarchies in their policies.

In this new system, the main difficulty is coordinating public policies within the central administrative network, within the political network, and between the two networks.

The development of cooperation among communes

The coordination of public policies in city regions has occurred through the development of cooperation among communes. The first big attempt to develop such cooperation was in 1966, when 'urban communities' were created in some large cities (such as Lille and Lyon)[13]. The next important attempts occurred in 1992 and 1999, when new intercommunal institutions (*communautés de communes*, *communautés d'agglomération* and *communautés urbaines*) were set up with good success, particularly in the Bouches du Rhône *département*, where the Marseille-Aix city region is located.

Institutional and spatial coordination at the metropolitan level has been linked in France to the decentralization national reforms. Local configurations follow the characteristics of the local institutional cultures, and coordination is very different from one metropole to another (Motte (1997)).

Decentralization during the 80s was organized with no hierarchy between the main levels of local and regional governments. *Communes*, *départements* and *régions* could act separately, and they were formally independent. The central state itself relinquished many elements of its hierarchical powers over local authorities. The existence of horizontal relations and coordination of agendas depended then of the willingness of the local governments (Thoenig and Duran (1996)). During the 80s, in some urban areas, this new situation led to the development of cooperation (Lille, Lyon and Rennes for example). In other urban areas, limited cooperation or non-cooperation was dominant (Bordeaux, Grenoble, Toulouse and Nantes for example). In the Marseille-Aix city region non-cooperation was the main characteristic of the institutional relations, and competition developed between *communes*.

The 90s period was different. With the results of the 1990 census, the central state discovered new growth within the Parisian region, and then launched national incentive policies to expand regional metropoles in the context of competition among European cities.

The Case of Marseille-Aix City Region

In the Marseille-Aix city region, new institutions appeared in 1992-1993, particularly around Marseille and Aix-en-Provence. They developed progressively and the 1999 national reform of the intercommunal structure gave birth to an 'urban community' around Marseille (Urban Community of Marseille Provence Métropole, consisting of 16 *communes*), and an 'agglomeration community' around Aix-en-Provence (Agglomeration Community of Pays d'Aix-en-Provence, consisting of 33 *communes*). Spatial

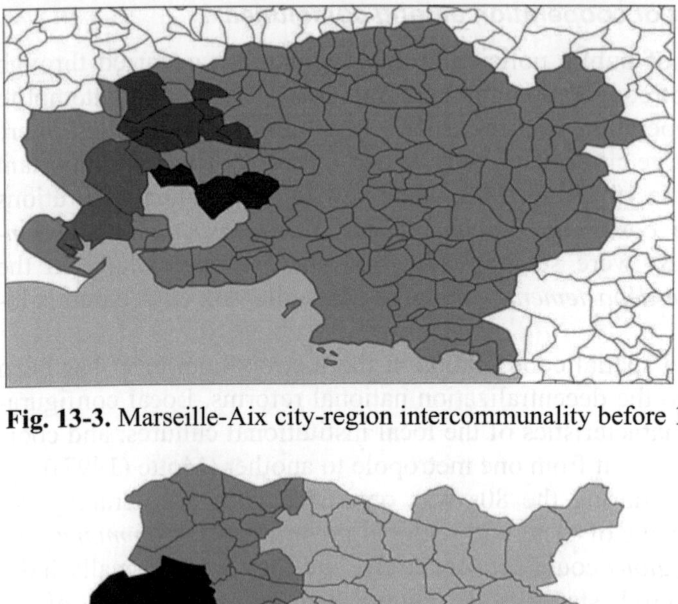

Fig. 13-3. Marseille-Aix city-region intercommunality before 1992

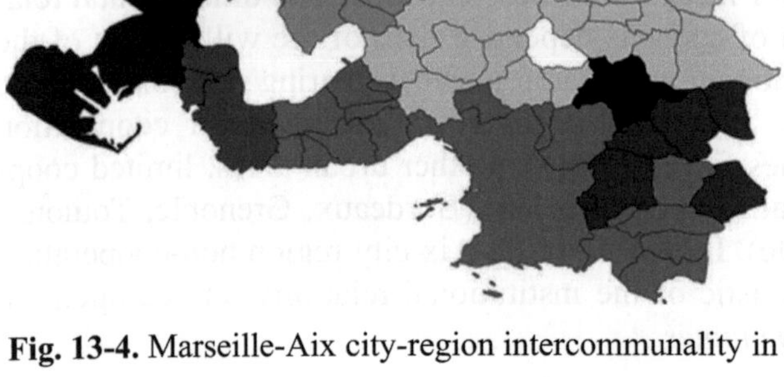

Fig. 13-4. Marseille-Aix city-region intercommunality in 2006

planning is one of the main tasks of those new institutions, and the new 2000 Planning Law has made spatial planning an important issue at the local level.

The central state played a crucial role during the 1980-2000 period in structuring and modernizing the institutions of this city region, first through national reforms and second through restructuring its administrative practices. At the end of the 90s the central state's position is explicit: "if we succeeded in structuring this *département* with 3 or 4 [intercommunal] institutions, for big issues of the agglomeration, we could then easily find the appropriate structures."[14] However, there was no city-region authority in place during the period between 1980 and 2000. Nevertheless, there is now a strong possibility for public policy coordination between the Marseille Provence Métropole Urban Community and the Aix-en-Provence Agglomeration Community, reinforced in 2001 by the election of mayors who belong to the same political sphere (right wing).

At the level of the Bouches du Rhône *département*, the central state developed actions to promote the building of new supra-communal institutions. The objective was to coordinate the production of public goods and services at the city-region scale, in order to scale up the competitiveness of the Marseille-Aix city region.

The political heads at the *commune*, *département*, and *région* levels are also important for understanding the Marseille-Aix city region. In Marseille, Socialist Robert Vigouroux was mayor from 1986 to 1995, but a shift to the right developed and Jean-Claude Gaudin (Démocratie Libérale - Right) was elected in 1995 and re-elected in 2001. At the département level, the Bouches du Rhône département remained with Socialist leadership (1986-2001). Despite the alignment of the political parties from 1986 to 1995, the département did not have a strong link with Marseille commune. In terms of the région level, in 1998 a Socialist president was elected as head of the Provence Alpes Côte d'Azur région, but his power is weak since he lacks a majority in the regional council.

Marseille's Socialist mayor Vigouroux had an affirmative policy for Marseille, trying to dominate, without much success, other *communes* of the *département*. He started nevertheless to cooperate with other communes around Marseille[15], and it was Vigouroux who created the intercommunal Marseille Provence Métropole in 1993 with 13 *communes*. His successor, mayor Jean-Claude Gaudin, developed a strategic spatial scheme for Marseille, and on a consensus basis expanded the Marseille Provence Métropole perimeter to include 20 *communes* in 1999. In 2000[16] he transformed it into a new 'urban community' (18 *communes*), which has strong powers in planning, economic development, housing, environment, and infrastructure provisions (water, sewage, transportation). In this process, mayor Gaudin tried to persuade the Aix-en-Provence mayor to join the Marseille Provence Métropole Urban Community, without success. One should note that Marseille Provence Métropole did not develop much during the 1993-1999 period. Policies were developed with an informal veto principle from the small *communes*, which prevented any large projects from emerging.

Aix-en-Provence used the 1992 and 1999[17] laws to strongly develop cooperation between *communes*. The Pays d'Aix Communes Community was created in 1993 (6 *communes*) and expanded to 19 communes in 1999. The Agglomeration Community was created in 2001 (33 *communes*), and has strong powers in planning, economic development, housing and environment.

Other Agglomeration Communities developed in 2001: Aubagne-Garlaban, Nord de l'Etang de Berre, Ouest de l'Etang de Berre, Nord-Ouest de l'Etang de Berre and Arles.

13.1.3 Issues and Conflicts on Large Urban Projects

Competition among the main public institutions (the *communes*, Bouches du Rhône *département* and Provence Alpes Côte d'Azur *région*) has nevertheless been the main characteristic of the 80s and most of the 90s, as seen in these metropolitan projects:

- Arbois projects (TGV rail station, road infrastructure development, activity zone development): conflicts between Aix, Marseille and the Bouches du Rhône département;
- Euroméditerranée project (311 hectares of urban redevelopment in the Marseille harbor area): not welcomed at first by the Bouches du Rhône département;
- Logistic: 3 projects were developed separately without much cooperation;
- Commercial zones: Grand Littoral development in Marseille was set up autonomously by the Marseille (Vigouroux) municipality.

A clear illustration of this lack of cooperation during the mid-90s is given by the case of large music halls built by Marseille (Le Dôme - 100MF) and Vitrolles (Le Stadium - 70MF). As of 2001 the Dôme was working well and the Stadium was collapsing.

Despite this competition, new cooperation emerged at the end of the 90s and the beginning of the century. Why such a dramatic evolution?

13.2 Spatial Strategic Planning for the Marseille-Aix City Region

A metropolitan strategic policy was defined by local representatives of the central state in 1969-1971 with the Schéma d’Aménagement de l’Aire Métropolitaine Marseillaise (1970)[18].

During the first period of the decentralization process, between 1981-1990, no metropolitan initiatives were taken by the regional or local authorities, or the central state.

A second period started in 1990. The central state played a very important incentive role in the emergence of metropolitan policy, a role which was developed both at the national level (new laws and policies) and in the Marseille-Aix city region. All the initiatives came from the central government, with some directness at times.

13.2.1 Central State Initiatives during the 90s

The initiatives taken by central government during the 90s developed regularly through all the political evolutions of government at the national level.

- At the beginning of the 90s, the Delegation for Regional Planning and Action (DATAR), the national government’s regional policy agency, launched a specific local mission to make metropolitan strategic policy proposals (de Gaudemar; de Roo): the Aire Métroplitaine Marseillaise was again considered has a top priority for the Aménagement du Territoire policies. The critical focus was to promote the exchange function of Marseille.
- From 1990 to 1999 the Exchange Club of the Metropolitan Area was set up by the Préfecture de Région. It was an informal network of metropolitan area planning practitioners, who generally met bimonthly, and argued their viewpoints on the main metropolitan issues, without referring formally to their institutional or political membership. They published numerous reports and two books in 1994[19] and 2000 (Langevin and Chouraqui (2000)). They analyzed and debated the metropolitan economic, social and spatial issues and collectively challenged local authorities about their inability to develop adequate policies to cope with those issues.
- From 1991 to 1996 the Ministry of Equipment (DDE, Direction Départementale de l’Equipement) proceeded more formally, as in

other metropolitan areas, through a strategic planning exercise (Brassart (ed) (1996)). The central state role, and particularly the DDE role, was to foster reflection on urban issues and develop shared points of view among public planning officers. Many meetings were organised with all the 'partners' of the Equipment Ministry (Region Prefecture, urban agencies of Aix-en-Provence, Marseille, DATAR, Agricultural Ministry, etc.)[20]. This 'partnership' was very broad and developed independently of elected representatives. It is a good illustration of the way the central state is proceeding, developing a common representation of the public necessity to act to face the stakes of the agglomeration. This planning exercise was concluded in 1996 with a book published locally by the Director of the DDE of Bouches du Rhône. His attitude was very critical of local authorities. He said for example, "The metropolitan area remains a virgin of any long term planning." Why? Because "there is historic hostility of this territory to any form of planning . . . This situation is at risk: development is incoherent and there are the perverse risks of laisser-faire."(Brassart (ed) (1996))

- In 1996, the implementation of the Pasqua Planning law of 1995 led to a Directive Territoriale d'Aménagement (DTA) procedure. In that new procedure, the central state could unilaterally define spatial policies for a territory. The only condition was that it had to be defined in consultation with the elected representatives. Since 1996, the préfecture and the DDE have elaborated this DTA at the metropolitan level, with many political and administrative difficulties. It was nevertheless a possible way to define metropolitan strategic policy. Jean-Paul Proust, regional *préfet*, said in 1999, "this document will give us a coherent organization for the whole metropolitan area."[21]

All those initiatives are emerging from the central state sphere: DATAR, ministries and their local representatives. Local initiatives outside this central state sphere are rare, particularly initiatives from the employers' community. The employers' community is in fact fragmented (Zalio (1999)), like the city region itself. That heterogeneity is the consequence of the decline in traditional economic activities since the 60s, and the arrival of large national and international enterprises during the same period. These large enterprises do not structure the local employers' community, and they coexisting with the traditional employers' communities. The heterogeneity is also linked to the fact that they benefit from the *communes*' competition to attract enterprises at any cost. This leads *communes* to supersede economic interests to all other interest.

13.2.2 Building the *Communautés* and the Elaboration of Development Plans (1999-2006)

The end of the 1990s marked another new phase. The convergence of national reforms both in institutional terms and strategic planning terms led to a completely new situation. The implementation of the 1999 Sustainable Development law makes compulsory the elaboration of a strategic spatial policy (*Projet d'agglomération*) for the urban area. The DATAR set up an agglomeration working group[22] between the Marseille and Aix communities to start the formation of an agglomeration strategic policy. This attempt was not successful but structured the emergence of political debate on this issue.

At the same time, the *région* had a new role during the 1998-2001 period because of the 2000-2006 Regional-National Plan Contract and the Regional Planning Scheme (*schéma d'aménagement régional*). The contract has some elements of a new city-region policy. It is also linked to National Service Schemes and European Programmes for the same period, and shows the links which are emerging among four levels of policies: Europe, France, region, and local communities.

The attitude of the central state was openly defined by *préfet* Jean-Paul Proust in 1999, clearly showing the new role of the central state: "My ambition, as a representative of the state, is certainly not to say what has to be done, but to try to federate all the energies to create a large project, an ambitious project for the metropolitan area." The central state effort has been on the institutional side, which could support the emergence of a city-region strategic policy.

What is new at the beginning of the century is that communes have started to experiment with some new cooperative relationships. Why is there such an emergence of cooperation among *communes*, after such a long period of competition?

Is this due to government policy in place since the beginning of the 1990s and to the considerable financial incentives in the 1999 law[23] which grouped the communes together? This factor is certainly important, but alone it does not suffice, since cooperation is much more limited in some neighbouring *départements* (Var and Hérault).

Is this linked to the fact that the councillors in communes on the outskirts of cities have become aware of the need to offer their citizens a greater standard of amenities and services than that possible according to the actual means of these *communes*? This factor must certainly play a role, for many outlying *communes* are not able alone to take on, for example, the cost of public transport or the collection and processing of household waste. Since the 1960s these services have been organized by resorting to forms of syndicate cooperation, but the boundaries of the syndicates vary from service to service. By the end of the 1990s the network of syndicate boundaries had become inextricable. In this respect the 1999 "Chevènement" law reforming the intercommunal institutions provided an opportunity to reduce the overly complex workings of local public services management, as well as the provision of amenities.

Is this linked to the chance hereby offered to mayors to be less isolated when dealing with the great complexity of the regulations that apply to the *communes*? Many municipalities are no longer capable of dealing with these regulations (for example, the application of European standards for environmental issues), which periodically leads to legal action against mayors. The intercommunal institutions, or *communauté*, allow management to be simplified, but the process involves unification of power which makes mayors very dependant on one another. Why have the mayors agreed to enter into this new reasoning that completely breaks with the isolationist traditions built up[24] after the end of the nineteenth century? In setting up a *communauté*, there are very important domains which fall under *communauté* authority (planning, economy, housing and social domains, at the least). The *communes* risk losing significant powers within these domains. The mayors agreed to enter because two new rules of the game appeared both in the Marseille Provence Métropole Urban Community and Aix-en-Provence Agglomeration Community: subsidiarity and consensus. This means firstly that each *commune* has broad leeway in choosing the policies which are really granted to the *communauté*. In order to transfer a policy to the *communauté* it must be declared to be of benefit to the *communauté*. This question of benefit has started to be framed by regulations, but it is interpreted by the *Conseil de communauté*. If this council works by consensus then the mayors obey the unwritten rule, namely they respect the choices of the other *municipalités*. The mayor thereby has considerable room to maneuver when it comes to policies that are to be transferred to the level of the *communauté*. To this level, conversely, he is able to transfer policies that he finds difficult to manage, those which pose awkward questions, or which he thinks are too costly such as access to public transport and the collection and processing of waste.

Did business leaders play a role in the incitement to group the *communes* together? We can consider their role to have been important. For several decades business leaders have been calling for a standard rate of local business taxation. The sum payable for businesses in the same sector can vary greatly from one *commune* to another depending on the abundance of businesses in the *commune* in which they are situated. This inequality between the rates distorts competition between businesses. The reform will allow the rates of local business taxation to be standardized across the *communauté* area and will in all likelihood allow the highest rates to be reduced.

These institutional developments could contribute to the emergence of a city-region spatial strategy elaborated by the local authorities. The central state's intense local activity during the 90s identified the main problems which have to be solved. Consequently, new appropriate city-region policies have been identified. The *communes* working together in new inter-communal institutions provide a strong possibility of coordinating these new city-region policies.

13.2.3 The Metropolitan Project Elaboration (2004 - 2006)

In June 2004 the French government (DATAR) issued a call for projects to promote the development of key French metropolitan cities (except the Paris area), entitled: "For a European development of the French metropoles: A call to metropolitan cooperation." As stated in its objectives in December 2003, the call focuses territorial policies on a new scale of public action: the city region. The nature of the actions at this scale was an innovation. It was not a call to immediately formulate an investment policy, but rather for the major public actors of a city region to set up a process to develop international projects, particularly those with an international competitiveness dimension. This call to cooperation was remarkably well received by the local communities, as all the main *communautés* in France, located in urban areas with more than 500,000 inhabitants, responded.

The metropolitan project Marseille-Aix was one of the first six projects approved by the DATAR, on February 25, 2005. The speed of the Marseille-Aix response was astonishing from a national point of view because it was at odds with the generally accepted image of a city region bogged down in adverse and insurmountable institutional conflicts between Marseille the commercial city and Aix-en-Provence the aristocratic city. What is the basis for this new territorial cooperation?

The response in December 2004 to the DATAR invitation shows the central state's influence on city-region planning. The invitation to tender was an opportunity of cooperation seized by the elected officials of the Marseille and Aix communities, but the initiative also lies with the central state. What are the characteristics of this interaction between the central and territorial levels? What are the dynamics of this territorial project at the city-region scale?

The call for metropolitan projects allows a profound change in the relations between the *communautés* at the city region. The relations were initially structured around the two central *communautés*: the Marseille Provence Métropole Urban Community and the Aix-en-Provence Agglomeration Community. These two *communautés* tried to overcome their own institutional logic to face the issues of the city region's international competitiveness. At the state's request all the *communautés* in the Bouches du Rhône *département* have been involved in the project since summer 2005.

The steps taken by the Marseille Provence Métropole Urban Community and the Aix-en-Provence Agglomeration Community are rooted in developments since the nineties, when the political and administrative spheres of the city region attained a better comprehension of the main issues, and formulated many sectoral territorial projects from the *communauté* scale to the city-region level.

The call to metropolitan cooperation made it possible for the local authorities to define a strategy for development accessible to activities with an international dimension.

The strategy of the local actors rests on priority given to the accessibility of the urban area, positioning it "*like a node of East-West and North-South exchanges*"[25] in order to make "a pivot of the Latin arc."[26] The high-speed train[27] project for Nice is important in this context, as the Marseille-Aix pair proposed and obtained a route different from the route which had been presented by the French Railway Network[28]. They argued that the creation of a high-speed European network along the Mediterranean arc could not take hold without Marseille. In addition to the LGV project, the strategy of prioritizing accessibility also forms the basis for developing the Marignane airport, the Port Authority of Marseille (PAM) in the Fos area, and the extension of the Grenoble-Sisteron highway.

The other facet of the competitiveness strategy rests on the reception of international activities. This policy begins with the Marseille operation Euroméditerranée and continues with the International Thermonuclear Experimental Reactor (ITER) project in Cadarache, which will push limits of the city region even farther north. This project of 4.7 billion euros joined together the European Union, the USA, China, Russia, Japan and South Korea with the ambition to control nuclear fusion and to thus revolutionize energy production. The economic repercussions for the area are estimated at 2 billion euros in the 10 years it will take to build the site, with the creation of 8,000 jobs. This project of international dimension will have repercussions on all the operations of the city region.

The call to metropolitan cooperation coincided with the call for competitiveness projects launched alongside the interdepartmental committee of regional planning in September 2004. It is based on mobilizing companies, research units and universities around geographical areas or common sectoral projects. Their goal is to reinforce the industrial potential of

France and to generate high value-added activities to stop the flight of companies and the processes of delocalization.

The parallel between these two steps makes it possible to underline the partnership character of the new incentive policies of the central state. The reorganization of the modes of territorial governance seems a preamble essential to the international development of French cities in Europe. Moreover, it becomes the indispensable condition for economic excellence through the development of synergies between public and private actors.

13.3 Urban Regeneration: The Euromediterranée Case

Euroméditerranée is a large-scale reorganization of the port and urban zones located at the back of the Marseille harbor. Controlled by a national publicly owned establishment (EPAEM, Etablissement public d'aménagement Euroméditerranée), it is the first operation of national interest (OIN) to include in its perimeter (311 hectares) numerous inhabited districts: 28,700 inhabitants for 14,500 residences. The perimeter of Euroméditerranée gathers heterogeneous and degraded territories, but they are nevertheless very often still in use. Haussmannien 19th century buildings, sometimes derelict and nearly deserted, interlacing expressways, sheds and warehouses inherited from formerly flourishing harbor activity lie near the city center.

In crisis for nearly thirty years, Marseille has been marginalized within a prosperous, but completely parcelled out metropolitan surface (Langevin (2001)); (Langevin (2007)); (Olive and Oppenheim (2001)); (Morel (1999); (Motte (2007)). The exodus of the middle classes, deindustrialization, the decline of harbor activities, and the exhaustion of tertiary employment resulted in increasing unemployment and a significant fall in tax revenues. At the time of the last mandates of Gaston Defferre, mayor of Marseille from 1953 to 1986, the city did not see any large economic development projects, except for the unfinished Directional Center, with which Euroméditerranée did not completely break up. In 1986, the election of mayor Robert Vigouroux became an occasion to openly raise the question of the future of the city, at the time of the harbor crisis.

Thus, the challenge of Euroméditerranée is simultaneously to boost the local economy by attracting tertiary-sector companies to rehabilitate the old districts, to develop a wider social mix through private investment, and finally to generate "a new sense of appropriation by the Marseillais of their littoral geography and a larger city center." Therefore, as we can see, the challenge is not only to reinforce the economic role of Marseille, but also to make certain districts recover a social centrality through enhancing the functions of habitat and urbanity.

13.3.1 Governance Construction on a Territory in Social and Economic Crisis and the Emergence of a Regeneration Project

The Euroméditerranée project originated with the reflections of the managers of the Marseille Chamber of Commerce in the late 1980s, and not from a communal initiative. Indeed, at the time of his successive mandates, mayor Defferre rarely worked with the harbor sector, which he did not control. A key actor of this territory, the port authority (WFP), was perceived by the Mayor as an unverifiable coalition between the central administration, the Chamber of Commerce and Industry (CCI) and dockers' trade unions. It was mayor Robert Vigouroux who gradually changed this perception and reconsidered the relationship between the city and its harbor. In 1986, he appointed a former navy officer as adviser in charge of the harbor issue, and in June 1990 he instituted the first city-port technical commission in charge of imagining the future of this divided territory. But it is the Chamber of Commerce which published, on June 10, 1987, a report titled *Marseille Provence International*, a new ambition for the economy of Marseille and Provence looking toward 1992. This report carries the seeds of the restructuring policies for these districts and the idea of creating a tertiary center, while avoiding calling into question the place and the role of the port authority.

The principal trigger of this operation was a private initiative. The real estate group SARI, known for its role in the construction of la Defense in Paris, purchased the Docks in 1989. Certain location choices, not always coordinated, had indeed gradually reinforced the attraction of the area located behind the port: at the end of the 1980s, the Minister for the Sea stated that the port authority would maintain an important place in the administration of Marseille. At the same time, the French National Railway Company (SNCF) announced that Marseille would be connected to Paris by high-speed rail (TGV) arriving at the central railway station Saint-Charles. With the port on one side, the TGV on the other and with the rehabilitation of the Docks by a private promoter, the future of the district could then be considered with optimism after a decade of doubts. There remained the task of formalize the expectations of the different actors, a task civil engineers from the Ministry of the Equipment (DDE) started to do.

Encouraged by Robert Vigouroux, Alain Villaret, a civil engineer (IPC) and secretary-general of the city, helped the elected officials to think over the adjustment of this zone, together with another IPC engineer, Jean Frébault, then director of the DAU. The latter, in partnership with the DATAR, first of all contributed to work out a local request likely to be supported by the mayor of Marseille. Several working sessions, to which the local administrative and technical officers in charge contributed, led to a report intended for the mayor, containing a proposal for the creation of a publicly owned establishment, in connection with the operation Marseille Métropole Euroméditerranée.

However, to support the operation, the DDE lay down two conditions, under pressure from the Ministry of the Interior and the Ministry of Finance. First, the municipality should not be the only actor responsible for the project; rather, it should be supported by the *région* and the *département*. Second, the city and its partners should present a project for a Mediterranean Marseille Métropole tertiary center much more ambitious than what had been defended by the city. This project was presented at the town council in autumn 1991.

The Euroméditerranée project was thus born from the conjunction of local and national wills, and common concerns over the future of Marseille. The decision to create a publicly owned establishment illustrates these concerns. The mayor at the time, Robert Vigouroux, then his successor, Jean-Claude Gaudin, were conscious of the weaknesses of the city's resources. A publicly owned establishment was seen as a guarantee of the presence of the state in financing future projects. On the other hand, the state saw in a publicly owned establishment the means of ordering the confused play of the local authorities, considered unable to reach an agreement. In August 1993, the report submitted by the "Masson mission" came to the same conclusion: that an implication at the territorial levels was indispensable to the success of the project, speaking even on this subject about a "union sacrée." Ultimately, all the actors agreed to work together on the future of Marseille, but nobody knew what to do concretely: Euroméditerranée was first a great political project, more than a program of urban development.

13.3.2 Development of a Multi-dimensional Sustainable Urban Regeneration Project

At its early stage the project was economic. It became multi-dimensional during the 1990s. In 1994, the operation suggested by the state echoed the concerns of the city, and related to the constitution of a tertiary pole, dependent on the tradition of Marseille trade, covering 250 hectares in the zone of the port. Its objective was centered on the revival of the Marseille economy, without it being an urban question of town planning or regeneration. In the report published by the city in July-August 1994, Alain Masson assigned to the project the ambition "to make of Marseille an international trading center."

The conclusions of the "Masson mission," whose report was published in August 1993, marked a decisive stage in the genesis of Euroméditerranée. This report was capital, not because of its contents in terms of installation, which was criticized locally as well as in Paris, but because it fixed the terms of the negotiations to come. It presented in particular the first operation budget, and defined specifically the distribution of the contributions between the state and the communities, which has not changed since. It is the same document which delimited the space perimeter of Euroméditerranée, after long local discussions. The government then appointed a second civil engineer, Jean-Pierre Weiss, to direct a mission of prefiguration on the site. Out of his work, the Euroméditerranée development agency (EPAEM) was created on October 13, 1995.

The latter was built along a pattern of partnership, according to an innovative diagram. It associates the national government, which is its principal backer, with all the local authorities: *ville*, *département* and *région*, to which are added, in a thoroughly formal way, the Marseille Provence Métropole Urban Community. Euroméditerranée is an 'operation of national interest,' which means that it benefits from derivative legal status for the benefit of the central state, a form of positive discrimination for the territory.

However, the distribution of the capital testifies to the will of the central state not to assume only the financial, but also the political responsibility for the project. Rather than directing the operation, the central state, as in many other cases, tries to generate local dynamics and even new forms of cooperation between the different actors of the territory (Duran and Thoenig (1996)). Therefore Euroméditerranée should not be interpreted in terms of requirements to be implemented, but rather as a space of ceaseless negotiations for the constitution of a collective project among multiple actors, whose participation is not fixed in time.

There is not one, but several pilots within the project. Although in theory the EPAEM is the main actor in the decision-making process, in practice it must act jointly with its local partners, while at the same time being assessed as the only actor in charge of the operations carried out on the site. The situation is all the more complex as EPAEM does not control all the relevant urban planning procedures, because they lie outside either its administrative or territorial jurisdiction. This dependence, moreover, is reinforced by human resource weaknesses. In spite of the heterogeneity of its tasks, the EPAEM is composed of a team of only about 50 people and is thus constrained, in its daily work, to rely on the technical expertise of the municipality, and even the external services of the central state. This institutional arrangement results in of the need for on-going negotiations regarding financial expenses and the technical responsibilities shared among EPAEM, the local communities and the external services provided by the central state. Although the steering committees constitute effective tools for administrative coordination, the project very often appears as a tangle of more or less autonomous devices.

13.3.3 Attracting Private Actors to a Large Project

The challenge of Euroméditerranée is also to attract the various actors, whether they be institutional or private, into a dynamic process to make them real partners. The difficult conditions of the site, often derelict as described above, the multiple actors, and the resulting tensions force the developers to negotiate step by step with multiple partners, often involving complex procedures and financial arrangements. Two examples perfectly illustrate this daily work of negotiations: the creation of a new business district in la Joliette, which is successfully underway, and the rehabilitation of the Rue de la République, although this programme is difficult to initiate. They also underscore the ambivalences of urban regeneration founded primarily on private investment.

The ZAC Joliette: A New Business District

On 22 hectares, the project is to create an international business district starting from an operation already carried out: the building of the Docks. This entails the creation or the rehabilitation of 300,000 square meters of buildings, two-thirds of which will be devoted to the tertiary sector and office buildings, the remainder having to accommodate some residences and a certain number of services (hotels and shops). The program for public infrastructure is also extremely ambitious, as the EPAEM envisages the creation of more than 1,000 parking bays, an underground network for part of the motor vehicle traffic, including one penetrating highway, and the creation of a great central square of several hectares. The ZAC Joliette offers a particularly significant example of public-private co-definition of a town planning operation, after SARI showed the way by rehabilitating the old Docks.

An important collective task related to the definition of the types of products to be offered to the companies. Marseilles had few office buildings of international class likely to accommodate the most demanding companies. The ZAC Joliette has succeeded in enabling the town to obtain products which had long been missing. This is particularly true concerning the buildings accommodating telecom and internet operators, which require high-quality protected sites with perfect connections to the electricity supply and fiber-optic networks.

Finally, a significant part of EPAEM's work consisted of accompanying the promoters interested in their research by investors upstream and customers downstream, in order to show the relevance of its visions.

In such a context, EPAEM's role is not necessarily to ensure the continuity of a project which is not fixed in time, but rather to ensure the coordination of the various initiatives aimed at making this part of the city recover its centrality. Day after day, EPAEM tries to reconcile the immediate logic of the promoters and its longer-term objectives. It was the case when, wishing to develop an offer of residences at ZAC Joliette, it worked for several years trying to convince a promoter to develop this product after it had agreed to purchase the land in order to facilitate his task. Another example is the small housing project carried out by George V, which proved to be a commercial success, thanks to innovative sales engineering.

The Rue de la République: A Private Operation with Public Initiative

The Rue de la République, a long avenue of Hausmannian architecture inhabited by poor residents, occupies a special place within the Euroméditerranée. Two features make it different from the rest of the operation. First, the private character of the buildings makes this the greatest private rehabilitation project in Europe. Second, the significant roles are played two big landowners: Eurazeo, related to the Parisian banker Lazard, and, recently, Lone Star, an international company dealing with investment funds. Generally, in Marseille about 50% of real estate belongs to small investors. However, here more than 75% is in the hands of companies connected to foreign industrial and financial groups which are, at least in one case, little sensitized to the overall project. The developer's challenge is to entice these private actors to public action, as for them the Rue de la République is private property before being an urban problem.

Another exceptional aspect deserves attention: the vacancy rates and the state of dilapidation, particularly serious for central housing, since 30 to 50% of the park can be regarded as derelict. The vacancy rates for the residences lie between 16% and 33%, depending on the building. The situation is similar for the shops and offices. The vacancy rates for the offices is almost 40%, and an average of 17% for shops. For many years, the street has been partially derelict, as the owners have not done any maintenance.

Rehabilitating an urban sector in crisis requires seizing the tensions generated by the interactions on several levels, the uncertainties which burden the behavior of the actors, and the procedures, more or less formal, that they implement to ensure the safety of their transactions, to establish bonds of confidence and to share the risks of possible failure. It is a process whose unfolding and objectives are concealed with anticipations, and is definitely quite difficult for a public body. This is all the more difficult as the engagement of the investors is not simply related to the confidence in the success of the operation brought to them by the presence of the central state, but also depends on the financial arbitrations operated by others, following logics which are foreign to the construction of this collective good which is the future of the city (financial health of the group, conversion of the real estate resources, pressure of the shareholders, mergers, etc.). All this creates an impression of groping public action. It is difficult to say if it marks a new form of public intervention, oriented towards the security of long-term decisions and engagements in a universe dominated by the short term.

13.4 Reflections and Questions about Spatial Strategic Planning and Large Urban Projects

The main questions raised by the planning context we have been studying relate to the nature of city-region planning. What is the shape it must or it can take? Should it have to integrate the large urban projects which are in practice one of its structuring dimensions? Indeed, the rational and logical answer can only be positive. Theoretically it is highly desirable that a planning policy should be set up at the level of the city region, in order to organize development of the urbanized territory. Planning at the scale of the urban area poses development terms that should prove socially balanced, economically effective and respectful of the natural resources. It is essential that the large projects which structure the city region should be developed from an integrative perspective to optimize public investments, support business management, and make housing conditions as comfortable as possible for residents.

There is little likehood of such planning in the Marseille-Aix city region, and I think that it seldom happens elsewhere, if one observes carefully the practices and not only the plans created by the planning processes. In fact it does not correspond to the dominant practices of the local authorities in France and Europe. Why? Are the necessary instruments for planning at the city-region scale missing? Are the bureaucrats and specialists insufficiently qualified, either administratively or technically? The instruments of planning exist, together with, generally, competent bureaucrats and technicians. The answer is to be found with local elected officials who, in a given administrative and technical territorial structure, which is generally institutionally split, have to merge the interests of their citizens towards common development objectives. However, the interests of the different social groups, which are politically and institutionally organized at the commune level, are in competition for the localization of what is the most desirable: high-tech non-polluting companies and middle-class or upper-middle-class households.

So, to overcome these contradictory interests, the elected representatives have to converge towards the same collective action values in a globally integrated planning process.

The first point is the economic competitiveness of the city region, through supporting the national and international companies in the urban area. This value is to be found first and foremost in the most advanced experiments of planning, and enabled the creation and development of the Euroméditerranée project. It is also found in the ITER project and the "competitive poles." Mobility, and thus the transport and communication infrastructure, partake logically in this search for efficiency and is at the basis of cooperation between elected officials.

The second point lies in fundamental political values such as social and territorial equity and social diversity, which are linked to the settlement strategies of the city region. Initial compromises on these values is very difficult in this context, and they will not emerge easily the Marseille-Aix city region.

The third aspect is the well-balanced management of natural resources, particularly in terms of water resources, waste management, and air quality. There are often conflicting projects between elected officials, who find it difficult to reach a compromise. The recent conflicts about waste processing of the Marseille Provence Métropole Urban Community and the Nord-Ouest de l'Etang de Berre (Fos) community illustrate how it is quite impossible to reach an agreement. This situation lead to authoritarian decisions by the central state.

Large projects are developed on partial dimensions of the city region. Therefore, only planning processes for specific issues can exist, and not an integrated planning procedure. Urban planners have to think over the situation as it is and not elaborate an a priori model of action that does not correspond to policy and social territorial realities. The planners must work to develop new visions of the city region in dealing with the specific issues. The combination of these visions through debates among the actors could help foster common views about the city-region development. In this way, it will become possible to increase the efficiency of collective action in the Marseilles-Aix city region.

Notes

[1] Parts of the paper (Outline, Strategic planning) were published in Salet, W., Kreukels, T., Thornley, A. (ed.) (2003) *Metropolitan Governance and Spatial Planning in Europe*. Spon. See pages 320-336. Evolutions between 2003 and 2007 have been introduced in this chapter.

[2] During the 60s 'urban communities' including a large number of municipalities were legally imposed by the government in many big cities, including Lille and Lyon. Marseille rejected the project of the urban community.

[3] Marseille is the most important *commune* of the Marseille-Aix city-region area with 57% of the total population, and 43% of the Bouches du Rhône *département*'s population.

[4] Those *communes* have very large territories, which is specific to that part of France.

[5] Statistics from the Bouches du Rhône *département*. During the 38 years from 1962 to 1999, there was 46.9% growth.

[6] This is partly due to the fact that these *communes* have large territories, which is an exception in France.

[7] *Centre d'Études Techniques du ministère de l'Équipement* --- Research Center of the Public Works Ministry

[8] *Centre d'Études Atomiques* --- Nuclear energy research center

[9] Residential development has also been reinforced by the massive investments in water supply by the central government during the 60s and 70s.

[10] Paris airports (73.6 million passengers in 2000), Nice (9.3 million) and Lyon (6.02 million).

[11] Département elected body.

[12] Elected body.

[13] And not in Marseille.

[14] *Préfet de Région* declaration In Langevin, Chouraqui, 2000, p. 182

[15] Using the 1992 Law

[16] Using the 1999 « Chevènement » law

[17] Loi « Chevènement » : Loi n° 99-586 du 12 juillet 1999 (J.O. du 13 juillet 1999, pp. 10361-10396)

[18] Planning Scheme of the Marseille Metropolitan Area

[19] Club de l'AMM : « la métropole inachevée »

[20] DIREN, DRE, CETE, EPAREB, Euromed.

[21] 1999, April, 30th

[22] The same experience started in 15 other French agglomerations.

[23] « Chevènement » Law : Law N°. 99-586, 12th July 1999 (J.O. 13th July 1999)

[24] See Crozier and Friedberg (1977) for a detailed description of the characteristics of the « système politico-administratif départemental français », 218-234

[25] AGAM (2005) *Métropole*, la lettre d'information de l'AGAM, p2

[26] AGAM (2005) *Métropole*, la lettre d'information de l'AGAM, p2

[27] Ligne Grand Vitesse (LGV)

[28] Réseau Ferré de France (RFF) better known as SNCF

References

Becquart, D. (1994) "Marseille : 25 ans de planification urbaine", *La Tour d'Aigues*

Brassart, E. (ed) (1996) "La région urbaine marseillaise", *La Tour d'Aigues*.

Club d'échanges et de réflexions sur l'aire métropolitaine marseillaise (1994) "La métropole inachevée Les ferments d'une démarche de prospective partagée", *La Tour d'Aigues*

Dubois, J. and Olive, M. (2001) "Euroméditerranée : un grand projet d'aménagement à l'épreuve du débat public", in Donzel A. (ed) *Métropolisation, gouvernance et citoyenneté dans la région urbaine marseillaise*. Maisonneuve & Larose

Dubois, J. and Olive, M. (2003) "L'aménagement urbain dans son horizon d'incertitude", in Fournier, P. and Mazzella, S. (ed) *op. cit.*

Dubois, J. and Olive, M. (2004) "Négocier la requalification urbaine. L'action publique dans son horizon d'incertitude", in Fournier, P. and Mazzella, S. (ed) *op. cit.*

Dubois, J. (2006) "L'espace normé ne fait pas l'espace du projet. L'opération d'intérêt national EuroMéditerranée de Marseille", in Bourdin et al. (eds) *Les règles du jeu urbain*. Descartes

Donzel, A. (2001) "Métropolisation, gouvernance et citoyenneté dans la région urbaine marseillaise", *Paris*

Duran, P. and Thoenig, J.C. (1996) "L'Etat et la gestion publique territoriale", *Revue Française de Science Politique*, 1996

Fournier, P. and Mazzella, S. (eds) (2004) *Marseille, entre ville et ports. Les destins de la rue de la République*, La Découverte, coll. Recherches

Healey, P., Khakee, A., Motte, A. and Needham, B. (1997) *Making Strategic Plans: Innovations in Spatial Plan Making in Europe*. University College Press Limited

Healey, P. (2007) *Urban complexity and Spatial Strategies Toward a relational planning for our times*. Routledge

I.N.S.E.E.-PACA (2002) *Atlas des métropolitains de la région urbaine de Marseille-Aix-en-Provence*

Langevin, P. and Chouraqui, E. (2000) "Aire métropolitaine marseillaise encore un effort...", *La Tour d'Aigues*

Langevin, P. and Juan, J.C. (eds) (2007) *Marseille une métropole entre Europe et Méditerranée*. La Documentation Française

Ministère de l'Equipement and O.R.E.A.M. (1970) *Schéma d'aménagement de l'aire métropolitaine marseillaise*

Morel, B. (1999) *Marseille, naissance d'une métropole*. L'Harmattan

Motte, A. (ed) (1995) *Schéma directeur et Projet d'agglomération: Un renouvellement du mode de gestion des espaces urbanisés français (1981 - 1993)*. Juris Service éditions

Motte, A. (2006) *La notion de Planification stratégique spatialisée. 2006*. Ministère de l'Équipement (PUCA)

Motte, A., Da Silva, J. and Douay, N. (2007) "La genèse d'un projet métropolitain improbable" in *Les agglomérations françaises face aux défis métropolitains*

Motte, A. (ed) (2007) *Les agglomérations françaises face aux défis métropolitains*. Economica/Anthropos (Collection Villes)

Salet, W., Kreukels, T. and Thornley, A. (eds) (2003) *Metropolitan Governance and Spatial Planning in Europe*. Spon

Zalio, P.P. (1999) *Grandes familles de Marseille au XXème siècle. Enquête sur l'identité économique d'un territoire portuaire*. Socio-Histoires

Part V

AMERICA

14. Regional Planning and Local Governance: The Portland Story

Ethan Paul Seltzer

14.1 Introduction

The Portland, Oregon, metropolitan region has acquired an international reputation for regional planning and governance. Planners, designers, and civic leaders from around the world have visited the Portland region as a means for gathering information about what it means to plan and govern at a metropolitan scale. Why this interest in regional planning and governance? Why Portland?

First, metropolitan regions have become the unit for analysis and strategic action in an increasingly global economy. This is the era of the modern "citistate," where national economies and economic competitiveness are assessed based on the health and strength of metropolitan economies. Second, Portland is of interest because, unlike any other metropolitan region in the United States, it has a regional government whose legislators are directly elected by the citizens of the metropolitan area. Furthermore, for over 30 years, it has been planning and acting at a metropolitan scale, and, unlike almost every other metropolitan region in North America, has a collection of projects, interventions, and plans, with metropolitan-scale outcomes.

This paper will present the planning experience of the Portland region, and the outcomes of the most recent series of regional plans and actions. This is a plan-led rather than project-based experience, and for that reason it is somewhat unique. Though Portland has benefited from a regional form of government, the degree to which the region has been successful depends in large part on the the embrace of a participatory strategy on the part of planners and decision makers, and implementation via broad partnerships able to span public, private and nonprofit sectors. This is a strategy available to all regions, whether governed by a single institution or not.

Though the Portland experience is useful, readers are cautioned to keep in mind that each place must make its own successes. The experiences and lessons related here are not provided as "best practices," to be copied faithfully in new settings. Perhaps the most important observation of the Portland experience is that trends can be changed. This story is not a prescription, but instead stands as an example of what can be accomplished if communities and regions seek to make changes in the trends that often appear to narrowly define options and outcomes.

14.2 The Oregon Planning Context

The Portland metropolitan region consists of five counties (Clackamas, Columbia, Multnomah, Washington, and Yamhill) located at the northern end of Oregon's Willamette River valley and one county (Clark) in southwestern Washington State. The current population totals about 2.1 million, with about 1.1 million jobs. It is a bistate metropolitan region, the only bistate region in the western United States, and encompasses 55 local units of government and over 130 limited purpose service districts.

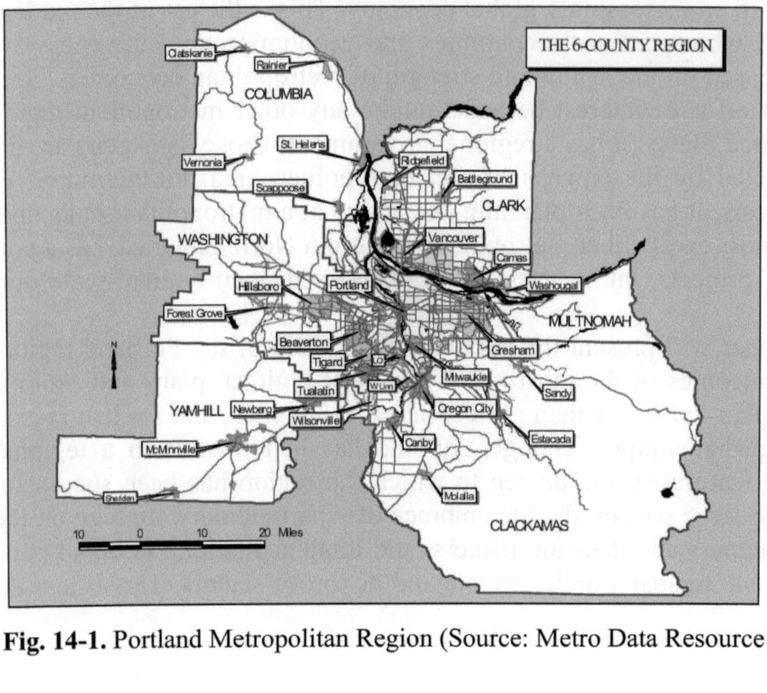

Fig. 14-1. Portland Metropolitan Region (Source: Metro Data Resource Center)

However, unlike other metropolitan regions in the United States, it is also home to a directly elected regional government, known as "Metro," in the Oregon portion of the metropolitan region. Metro will be discussed in the next section. This paper focuses on the period defined by the adoption of the Metro "home rule" Charter (Metro (2003a)), Regional Urban Growth Goals and Objectives (Metro (1991)), Region 2040 Growth Concept (Metro (1997)); (Metro (2000b)), Regional Framework Plan (Metro (1997)), Regional Land Information System (Knaap et al. (2003)), Regional Greenspaces Program (Metro (1997)); (Metro (2001)), and the Future Vision (Metro (1995)), all adopted or implemented from 1991-1997.

The context for these actions is provided by the Oregon Land Use Planning program which requires all cities and counties in Oregon to adopt comprehensive land use plans. Those plans must substantively address the Oregon Statewide Planning goals, and must demonstrate that they have done so via findings of fact.

In Oregon, despite a mandate to engage in comprehensive planning and a requirement that comprehensive plans serve, at a minimum, the statewide planning goals, there is no statewide land use plan. Rather, to the extent that there is a plan for the state, it is provided by the "quilt" of plans resulting from local planning action. Consequently, though the state requires planning and has the authority to reject a plan for inadequately meeting the statewide goals, planning in Oregon is ultimately a product of local action. In this regard, it is not wholly top-down or bottom-up, but a combination of both. It puts important power in local hands, but retains an important oversight role for the state.

Regional planning and growth management, therefore, takes place within a context of state-local relations established through the ongoing statewide land use planning program. Local comprehensive plans are specific, consistent with state and regional plans, and guiding when it comes to managing growth and change. Metro is not an implementer or empowered to develop a comprehensive plan. Rather, regional planning is intended to be coordinating and complementary for local planning efforts, and, at the same time, it is expected to provide a regional perspective on growth and change in the metropolitan area.

The statewide land use planning program has had its own share of controversy over the years, with three direct challenges at the polls (all defeated), and more recently the passage of a "property rights" measure (Oregon Measure 37) that potentially will cripple the planning program that has been in place now for over 34 years. Though statewide planning requirements, particularly those calling for urban growth boundaries around all cities and urban places in the state, are essential for making regional planning and growth management possible, they have also created institutional and political circumstances that frame and complicate regional efforts and regional-local inter-relationships.

Finally, at a very local level, the Portland region has what urban historian Carl Abbott calls an "intentional civic culture." Citizens expect to be directly involved in land use planning. The City of Portland, for example, has 95 neighborhood associations through which citizens have been directly involved in land use and other city issues for over 30 years. Most suburban jurisdictions have developed similar structures for community organization and interaction. This expectation for involvement extends to the regional level, not just in terms of inter-jurisdictional relations but directly to individual households and communities as well. Citizen involvement at a regional scale, along with the already complex inter-jurisdictional/inter-institutional set of relationships and legal requirements are all a part of the Portland story, to which we now turn.

14.3 Metro

Metro's origins can be traced to a grant-funded initiative in the mid-1970s to study and propose regional governance options for metropolitan areas (Abbott and Abbott (1991)); (Abbott (2001)). That project resulted in a report to the legislature, and the referral of a proposal to the voters to create Metro in the Portland region by merging the old, and out-of-favor, council of governments with a limited purpose regional service district. The voters approved the measure in 1978, and in 1979 the first Metro Council was elected.

In 1990, the voters approved a constitutional amendment to allow home rule charters for metropolitan governments in Oregon, and to enable existing metropolitan governments to appoint charter commissions. Metro appointed a charter committee, and in 1992 the voters within Metro's boundaries approved a home rule charter for the agency. Significantly, that charter:

1. reduced the number of Councilors from 12 to 7,
2. added an elected auditor,
3. established a standing system of policy advisory committees,
4. spelled out provisions for engaging in new areas of regional concern, and
5. most significantly, declared urban growth management to be Metro's most important task.

In 2000, the Metro Council referred a charter amendment to the voters

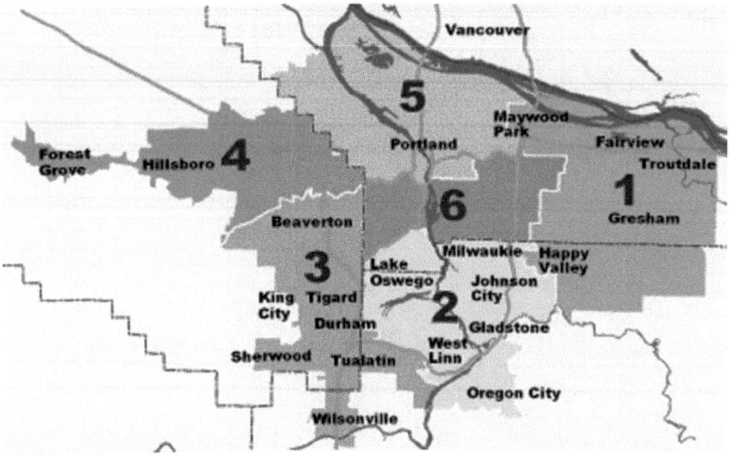

Fig. 14-2. Metro Council Districts (Source: Metro Data Resource Center)

to replace the elected Executive Officer with an appointed manager, decreasing the number of councilors from 7 to 6, and adding a Presiding Officer elected at-large.

Today, Metro has primary responsibility for regional land use, growth management, and transportation planning on the Oregon side of the Columbia River in the Portland-Vancouver metropolitan area. In addition, it is responsible for the management of the region's solid waste disposal system; regional convention, visitor and performance spaces; management and further development of a regional greenspaces system; and ongoing maintenance of a regional data and GIS system.

By state statute, Metro plays a unique role in land use planning. By law, only cities and counties have the responsibility and authority to develop comprehensive land use plans. Comprehensive land use plans, under the Oregon statewide land use planning program, are the primary legal document guiding all local planning and development decisions. By law in Oregon, zoning codes must implement comprehensive plans (Oregon Department of Land Conservation and Development (2001)).

Though Metro has not been given the authority to engage in comprehensive land use planning, it has been given the authority to develop regional functional plans. A functional plan addresses one or a narrow set of issues associated with an issue of regional significance. Furthermore, Metro has also been granted the power to require changes in local comprehensive plans to make them consistent with regional functional plans (Metro (2003a)). This is an astounding power for a regional agency in this country, more so because comprehensive plans are such central legal and policy documents for Oregon communities.

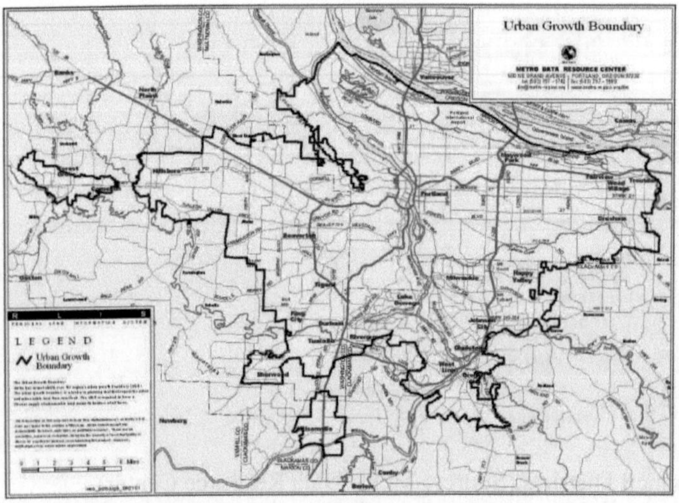

Fig. 14-3. Metro Urban Growth Boundary (Source: Metro Data Resource Center)

Significantly, the state has assigned to Metro the responsibility for establishing and managing an urban growth boundary (UGB) for the 24 cities and parts of three counties within its jurisdiction. In this instance, Metro furnishes a component of a comprehensive plan, here the UGB, to each of the jurisdictions. They are required to make their plans and implementing actions consistent with the UGB, and by doing so, are not required to show any other basis for asserting that their plan is consistent with state planning goals related to the creation of urban growth boundaries.

For the first decade of Metro's existence, the region was mired in a deep and prolonged recession, with little call for either new plans or expansion of the UGB. Most of Metro's regional growth management planning activity involved coordinating locally developed comprehensive plans with each other. However, the prospect, specter to some, of functional plans emanating from Metro loomed as a source of tension in the regional-local relationship. For most of its first decade, Metro's major planning activities focused on its role as the Metropolitan Planning Organization for Federal transportation planning and resource allocation purposes, and on the siting of landfills and management of the solid waste and recycling collection system.

14.4 RUGGO and the Region 2040 Concept

The conventional wisdom is that if you aren't growing, then nothing is happening. However, during the early 1980s in Portland, a time when the region's economy was in a deep and prolonged recession, a lot was happening. The metropolitan region lost both population and jobs in what turned out to be a major restructuring of the area's economy. Though the region was, by most accounts, declining, construction in the downtown core continued, a light rail system opened, and comprehensive land use plans that were prepared consistent with Oregon's statewide planning goals were adopted locally and approved by the state.

In the late 1980s, as the region's economy began to rebound, urban growth and its impacts on infrastructure, the landscape, and communities became an issue again. By late 1988 three issues began to dominate land use and policy discussions about the Metro urban growth boundary:

- Inside the urban growth boundary the region was experiencing many of the same issues associated with sprawl development elsewhere. Increasing traffic congestion , vehicle miles traveled per capita, and disappearing open space were becoming neighborhood, jurisdictional, and regional concerns.
- Outside, in rural areas, speculation was occurring everywhere, even on lands reserved for exclusive farm use. As the economy picked up, adjacent rural residential zones began to be occupied by high-end residential development. This posed a paradox to the region and the state: if the adjacent rural residential lands were built up with very expensive single family structures, then the use of those areas for future urban use, at urban densities, would be highly unlikely and future changes in the urban growth boundary would be forced to occur on the "vacant" resource land, the very land Oregon set out to protect in 1973.
- At the boundary itself there was no policy, no guidance for farmers, investors, elected officials, and local service providers as to where expansion might most likely take place.

In early 1989, the Metro Council adopted a workplan for periodic review of the urban growth boundary that called for the creation of Regional Urban Growth Goals and Objectives (RUGGO) as the underpinning for an overall regional urban growth management plan. This important step was not without controversy. For its first nine years of existence, Metro's commitment to land use planning amounted to a weak coordination function as local comprehensive plans were adopted, and by 1987 had dwindled to one half-time position charged with overseeing the boundary. Taking on a planning process like this demanded new commitment on the part of Metro to its land use and growth management planning role, and was accompanied by a growing regional growth management staff.

In addition, it raised the prospect of Metro exercising its functional planning powers, something it had never done in its first nine years. Local governments within Metro's jurisdiction knew of the functional planning authority, and made it clear that exercising that power would be regarded as a serious usurpation of local control. On the other hand, environmentalists, concerned by what they viewed as lax local concern with land use planning objectives, wanted Metro to step in and take action to counter local decisionmaking viewed as enabling sprawl-patterns of development.

Metro created the Urban Growth Management Policy Advisory Committee to oversee and propose goals and objectives for adoption by the Council. The Committee was chaired by Metro Councilor Jim Gardner and consisted of representatives from local government, the development community, environmental organizations, state and regional agencies, and citizens. It began its work in mid-1989 and in late 1991, the Regional Urban Growth Goals and Objectives were adopted (Metro (1991); Metro (1997)).

The goals and objectives accomplished two important tasks. First, as the result of a year-long negotiation, they spelled out for the first time how regional planning would be done, when Metro would exercise its considerable powers, and what the roles would be for other jurisdictions and interests in the regional planning process. Second, RUGGO served as the "sketchbook" for the region, providing a common framework for the growth management challenges of the day.

Parallel to the initiation of the periodic review effort, the staff in charge of Metro's data and information services started to create what is now the internationally recognized Regional Land Information System (RLIS). RLIS is a geographic information system that links a wide range of public records to a land parcel base map. Metro began developing RLIS in 1988; it was designed to be an urban planner's Geographic Information System (GIS), incorporating data essential for regional urban planning and growth management (Knaap et al. (2003)).

Designing RLIS was a collaborative effort, involving regional, county, and city planners. The objective was to identify the data and functional requirements of a GIS supporting community and regional planning. Its region-wide usage for planning and environmental management was to provide consistent land information across jurisdictional boundaries for GIS users in government and business, enabling data exchange and sharing of maintenance responsibilities.

In addition to developing a parcel-level base to support modeling and analysis, RLIS provided regional planners with a number of "firsts" essential for supporting existing and emerging regional planning efforts. It provided the first parcel-level base map for the region, showing the impact of the urban growth boundary in bold relief, regionwide, for the first time. It provided the first composite comprehensive plan and land use maps for the metropolitan region.

It provided the first data on rural and southwest Washington land cover and land use plans on the same maps showing the territory inside Metro's boundaries on the southern or Oregon side of the Columbia River. It enabled the development of a complete set of "McHargian" overlays needed to support environmental planning efforts linked to actual landscape characteristics. Finally, it provided the basis for creating new tools for citizen involvement, including real-time experiments showing the impacts of user-defined land use policy choices. Significantly, RLIS directly supported the complex land supply accounting needs required by state and regional expectations for the management of the urban growth boundary.

All three of these efforts --- RUGGO, the Metro Charter, and RLIS --- were developed through participatory processes that involved jurisdictions, citizens, business, and civic interests. Whereas Metro faced a large amount of skepticism, if not fear and opposition for expanding its regional planning role in the 1980's, by this time in the early 1990's a significant amount of trust had been built with key regional partners. In fact, it was suburban jurisdictions themselves that called for the next phase of regional planning, recognizing that the region's "sketchbook" needed further elaboration to serve as a vision. The Region 2040 Planning Project was developed specifically to add specific structure to the region's conception of its physical form. Through the Region 2040 planning process the region would:

1. specify the degree of expansion, if any, required of the urban growth boundary and the locations for any future expansions;
2. identify the major components for the regional transportation system, especially transit components and the creation of a regional pedestrian system;
3. identify a hierarchy and system of places, ranging from downtown Portland to existing town centers to regional centers and neighborhoods; and
4. incorporate a system of greenspaces in the urban region, both for purposes of accommodating outdoor recreation and for maintaining the viability of wildlife habitat.

The Region 2040 process was financially supported by Metro; Washington, Multnomah, and Clackamas counties; Tri-Met (the regional transit agency); and Portland General Electric. Representatives from each of these organizations constituted a management team that met regularly, sometimes weekly, to advise Metro on project strategy and content.

In addition, the Metropolitan Policy Advisory Committee, created as part of the Metro Charter, provided Metro with an advisory committee of local government officials, state agency heads, and citizen interests. The Metropolitan Technical Advisory Committee provided Metro with the advice of local government planners for both Region 2040 and other ongoing planning efforts. Finally, the Regional Citizens Involvement Coordinating Committee at Metro provided Region 2040 staff with advice regarding citizen outreach and involvement.

The first step in the Region 2040 project was to characterize base conditions (beginning with the history of settlement in the region), community values and expectations, and to create what became known as the "do nothing" or "base" case, the probable future if nothing was done and existing development patterns and dynamics went unchecked (Metro 2000b). The information on base conditions and community values was then used to propose three potential alternative urban form strategies for the region from which a preferred alternative could be constructed.

Alternative A called for less urban growth boundary expansion than the base case along with changes in the land use-transportation relationships along major corridors. Alternative B called for no expansion of the urban growth boundary, a large expansion of the transit system, and major rezoning. Alternative C called for minor changes in the urban growth boundary, a large expansion of the transit system, and the creation of satellite cities outside or separate from the current urban growth boundary.

Each of the alternatives and the base case were evaluated using a set of criteria developed through a public process. Extensive modeling of the transportation system, air quality, and land use allocations accompanied each scenario. Throughout this process, numerous public hearings and workshops were held, the project was publicized on cable TV and through the news media, over 25,000 newsletters were mailed to area households, and hundreds of presentations were made to local governments and civic organizations. Over 500,000 copies of a tabloid outlining the alternatives and the trade-offs involved in selecting different growth management strategies for the region were mailed to every household, resulting in over 17,000 citizen comments and suggestions.

Throughout the public involvement activities, Metro asked citizens to respond to four central growth management techniques: reducing average residential lot sizes, reducing parking, encouraging new growth to locate proximate to transit stations, and encouraging new growth to locate in existing city centers. Six other management techniques, ranging from the establishment of greenbelts to encouraging growth in neighboring cities, were also tested.

From the public involvement activities conducted over about a two-year period, Metro learned that participating citizens generally supported:

- Holding the current urban growth boundary in place.
- Utilizing the growth management "building blocks" tested throughout the process, especially establishing greenbelts and encouraging development in existing neighborhoods and close to transit.
- Reducing traffic and encouraging the use of alternative modes for travel.
- Retaining open space both inside and outside the urban growth boundary.
- A combination of increasing density inside the urban growth boundary and encouraging some growth in neighboring cities.
- Ongoing public education and dialogue regarding the trade-offs inherent in growth management.

Citizens questioned why growth had to occur, and whether planning to accommodate growth would only encourage it to happen. They were skeptical of using neighboring cities to accommodate growth, since they believed that it would be unlikely that enough jobs would be created in those locations to keep people employed close to home.

The preferred alternative resulted from a blend of public comment and aspects of each of the scenarios. What ultimately became the adopted 2040 regional growth concept called for very little urban growth boundary expansion over the next 50 years, amounting to about 7% of the existing total. These future expansions were targeted to parts of the region

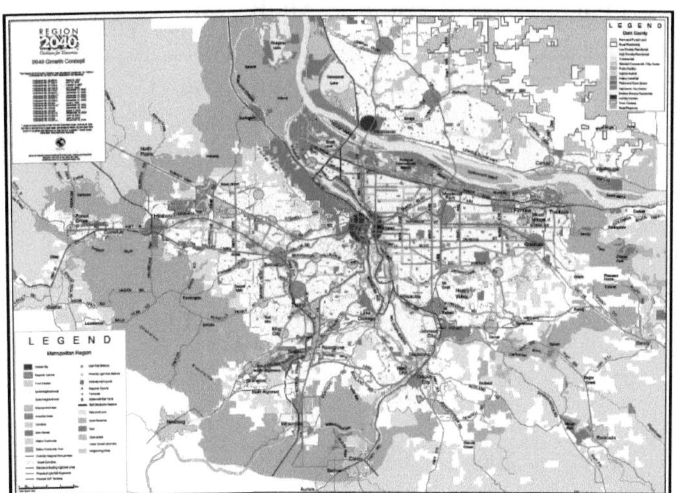

Fig. 14-4. 2040 Growth Concept (Source: Metro Data Resource Center)

needing additional urban growth but avoided the use of lands protected for farm and forest use.

In contrast, the base case and alternative A called for massive increases in the acreage devoted to urban use and expected very little in the way of redevelopment on existing urban land. Unlike alternative B, the adopted growth concept for 2040 calls for a minor amount of urban expansion, though it reflects an overall desire on the part of the public to limit urban expansion to the extent possible.

The transportation system concept called for additions to the light rail system backed up by higher expectations for pedestrian trips within and at each end of trips between important regional centers. Each center will likely serve different functions since this region cannot sustain a large number of places that offer the same things. Note that light rail is intended not so much as a replacement for the automobile, but as a device that allows people to inhabit the region as pedestrians.

A system of greenspaces was identified to separate communities from each other and continue to protect open space resources within the urban area. Finally, specific expectations were stated for the relationship between the urban area inside Metro's urban growth boundary and small rural communities outside the urban growth boundary.

Table 14-1 summarizes a number of the measurable differences between the alternatives and the preferred alternative as would be expected in the year 2040.

Table 14-1. Comparison of 2040 Alternative Scenarios (Source: Metro (2000b))

Attributes	1990	Base Case	Alt. A	Alt. B	Alt. C	Preferred Alt.
Single-family/Multi-Family (%)	70/30	70/30	74/26	60/40	69/31	65/35
% Growth Captured in 1990 UGB	100	83	71	100	63	87
% Growth via Redevelopment	---	0	6	18	8	19
Acres Farmland Converted	---	63,900	17,200	0	11,400	3,545
VMT/Capita	12.4	13.04	12.48	10.86	11.92	11.76
Mode Split (auto/transit/ped-bike)	92/3/5	92/3/5	91/4/5	88/6/6	89/5/6	88/6/6
Congested Road Miles	151	506	682	643	404	454
Transit Riders (1000's)	137	338	372	528	437	570

The preferred alternative and what became the final adopted Region 2040 Growth Concept received extensive public review using the same techniques outlined above. A video outlining the preferred alternative and its impact on the region was made available to the public for no charge through Blockbuster Video outlets, a chain of video rental stores.

Architect Peter Calthorpe, retained earlier in the process, developed a series of "regional design images", "before-and-after" elevations and site plans for locations throughout the region helpful for making the potential affects of the plan more tangible to a broader audience. In December of 1994, after two-and-a-half years, the Metro Council adopted the Metro 2040 Growth Concept at the urging of local governments, citizens, and business interests (Metro (2000b)).

14.5 Outcomes

As noted above, the Region 2040 Growth Concept created three major strategies for guiding the growth and development of the region:

- It identified areas for profound growth and change (a system of centers and specific areas for future Urban Growth Boundary amendments), leaving the rest of the region relatively unchanged.
- It developed a concept for the transportation system based on the expectation that it be multimodal and able to accommodate both radial and circumferential travel.
- It identified a system of greenspaces, parks, and trails to enable regional residents to live a life close to nature.

Even though the Regional Framework Plan has only been in place for approximately 10 years, it is possible to identify outcomes pertaining to each of these strategies.

14.5.1 Places for Growth and Change

Most urban growth has been effectively contained within the regional urban growth boundary. Despite the fact that population has been growing faster than anticipated, most of that growth has been captured within the urban growth boundary. Nonetheless, the need for new urban land to accommodate greater-than-expected growth has led to the first major expansion of the urban growth boundary onto lands identified in the Region 2040 planning process.

In December of 2002, the Metro Council took the unprecedented step of adding close to 18,000 acres to the Urban Growth Boundary, increasing its area by approximately 7% (Metro (2002)). Though small additions had been made through the years, this was the largest single legislative addition to the boundary in its history. The amendment was the result of state legislation requiring the maintenance of a 20-year supply of land for residential purposes.

However, in addition to new and rapid urban growth, the agricultural economy of the region has continued to grow. Between 1987 and 2002, the number of farms in the metropolitan region has increased, the number of farm acres has increased, and the net income per farm has increased 18%. Though this is the most urban part of the state of Oregon, it has also managed to retain a vibrant and dynamic agricultural sector, something not usually associated with fast-growing urban places in the United States.

Though the region had identified certain centers as key sites for promoting downtown-like function, this represented the first time that Metro's regional planning had specifically singled out a relatively narrow class of land uses and locations as being of regional significance. In a 2001 study Metro found that these centers were experiencing significant levels of underbuilding. That is, centers, though accommodating 58% of the region's employment on 21% of the land area, were not developing as rapidly as Metro planners has forecast. In essence, the zoning was and is ahead of the market. Several reasons were suggested:

- Most centers do not consist of greenfield sites and therefore require redevelopment while other potential sites for the same kind of development do not.
- In many cases, there are fewer regulations governing other commercial and industrial locations, making it easier to do the development desired in centers in alternative locations.
- To get the density desired in centers, parking needs to be accommodated either under ground or in parking structures, adding an extra and significant cost to development.

To overcome these barriers to center development at desired densities, Metro was advised that is needed to take several strategic actions. First, it needed to coordinate center development and zoning with development and zoning outside of centers. Simply put, there are too many noncenter locations that are too easy to develop, and there isn't enough overall market demand in the region to effectively raise the cost of land in a manner that makes center locations competitive. Second, Metro and local jurisdictions needed to carefully engage in regulatory review efforts to ensure that over-regulation in centers did not turn away desired development activity.

Third, Metro and local jurisdictions need to work closely with private interests in a public/private partnership to support pioneering center development through strategies like public participation in the funding of parking and other elements needed to accommodate and promote development at desired densities. In essence, centers are a useful aspiration, and important long-term goal, but their time has not yet come. Sustaining a long-term view of the role for centers in the growth and functioning of the region requires greater public participation in making that development competitive than was originally expected.

14.5.2 Transportation

Since its inception, Metro has been engaged in transportation planning aimed at developing a multimodal transportation system able to meet the wide range of needs for accessibility and mobility in the metropolitan area. Region 2040 provided new guidance for the incorporation of land use objectives into transportation system models, policy making, and planning. Further, Region 2040 provided Metro and the state with explicit guidance regarding the management of the urban growth boundary, something that prompted the creation of RUGGO and Region 2040 in the first place.

Today, the Portland metropolitan region is using its transportation system in a markedly different way than in most other metropolitan regions. Since 1996 the number of vehicle miles traveled per capita in single-occupant vehicles has been declining. The growth rate for transit boardings have outpaced both the rate of population growth and the overall rate of the growth of total vehicle miles traveled by a growing population. Approximately 70% of transit riders ride by choice, with nearly 60% leaving a car at home. Transit and non-auto travel among those located in the central city and centers significantly outpaces that for those located or working outside of centers.

All of this suggests that regional efforts to create a polycentric region have yielded positive results. The linkage of land use with transportation through both the Region 2040 efforts and other decisions made over the last 35 years appear to be well on the way to enabling regional residents to inhabit this place as pedestrians. The date support the notion that this is a transportation system designed to increase access rather than simply mobility.

14.5.3 Parks and Greenspaces

Parks and greenspaces, a signature structural element of the Regionl 2040 Growth Concept, also made a major advance in the wake of the regional planning effort. Beginning in the late 1980's, advocates for parks and natural areas worked with Metro staff to make the provision of greenspace an issue of regional concern. In 1992, the first Metropolitan Greenspaces Master Plan was adopted by the Metro Council. It outlined a vision for an interconnected system of regional parks and greenspaces able to meet needs for both habitat and recreation. The plan identified important sites as well as key links in a regional trails system, and implementation roles for citizens, local government, nonprofits, and businesses.

Then, concurrent with the development of the Region 2040 Growth Concept, parks and greenspaces advocates worked closely with Metro Councilors to bring forward a regionwide bond measure to raise $135.6 million to acquire key properties identified in the master plan. The greenspaces planning and campaign was spearheaded by greenspaces activist Mike Houck and a coalition of about 200 local advocacy groups known as FAUNA: Friends and Advocates of Urban Natural Areas. FAUNA provided the grassroots muscle in every corner of the region to support the levy, and the voters strongly supported its passage.

Since the approval of the levy by voters in May of 1995, Metro has acquired over 7915 acres, close to 2000 more than was originally thought to be possible (Metro (2001)). These sites and corridors are now permanently protected as part of the regional greenspaces system, itself an important structural element envisioned by Region 2040. In the fall of 2006 a second $227 million regional bond measure was put before the voters and it, too, passed with overwhelming support, thereby continuing public and governmental support for securing the land resources needed to sustain the urban population expected here in the future.

14.5.4 A Changing Context

Several important developments suggest that the context for regional planning and development is changing and will likely exert an important impact on the work of Metro and local jurisdictions. First, as was mentioned above, in 2004 the voters in Oregon passed an initiative that requires government to pay for any land value lost due to the imposition of new land use regulations, or not apply those regulations to the proposals of landowners being affected. Measure 37, as it is known, has created new ambiguity about the future for the Oregon statewide land use planning program and for urban growth boundaries in particular, an important part of what makes regional growth management in Oregon effective.

This ambiguity now about the future for land use planning in the state has led to the creation of an effort known as the "Big Look" at the state level, and a complimentary effort at the regional level known as the "New Look," to explore the challenges to and support for land use planning and growth management in coming decades. Public support remains high for land use planning and for protecting individual property rights. Reconciling and merging those two public objectives stand as a signature challenge to the future role for planning and intentional public intervention in the growth and change coming to Oregon and the metropolitan region.

Second, globalization has led to the identification of groups of potentially networked metropolitan regions in North America and elsewhere (www.America2050.org). In this instance, the Portland metropolitan region is envisioned as being part of the megaregion known as Cascadia, joining Portland with the metropolitan areas of Seattle, Washington, and Vancouver, British Columbia. Cascadia, along with nine other proposed US megaregions are expected to be the location for 70% of the nation's population and employment growth by 2030. Though the functional ramifications of megaregions are not yet known, this, too, may create a new set of forces that will shape and challenge regional planning efforts like the ones of recent decades in Portland.

14.6 Observations and Lessons Learned

What lessons, then can be gleaned from the Portland experience? Clearly, contextual factors have been profound in shaping the nature and timing of this experience. The history of planning and governance, the landscape itself, and the role of the Portland region on the west coast, among other factors, have exerted a strong impact on the planning outcomes experienced here. Nonetheless, there are several salient and transferable lessons that can be gleaned from this region's planning activities:

This is not an Experiment

Often the Portland experience is described as an experiment. It's not. This region did not set out to become a national model. Instead, it sought to serve the values that have consistently characterized this community: a real desire to make up our own minds and solve our own problems, and significant concern for the environment.

As Carl Abbott has noted, this is an intentional place, and what has happened is the result of an ingrained desire to serve it well (Abbott (2001)). As former Governor Tom McCall, "father" of the Oregon planning program, once said, "Heroes are not giant statues framed against a red sky. They are people who say: This is my community and it's my responsibility to make it better."

When it comes to the growth and change of any particular metropolitan area, there are no controls in the classic experimental sense. You can't go back and "run the experiment" again with a different set of parameters. Consequently, what has been taking place at the regional level in the Portland metropolitan area is not an experiment, just as what is happening in other communities, intentional or not, is no experiment either. Every choice made or avoided is part of what a place will become. More than most, the Portland region has chosen to make its future by choice, not by accident.

Planning Matters

Through planning, patterns that rule urban landscapes can be changed. As Lew Hopkins, Gerrit Knaap, and their colleagues have shown, plans inside the Portland urban growth boundary do matter (Knaap et al. (1999)). Prior to the creation of the urban growth boundary, the total number of new households that could be accommodated by the urban land supply was about 160,000. After adoption of the urban growth boundary, that same supply yielded about 310,000 potential new units. Further, the development of a more compact, efficient settlement pattern in combination with careful transportation planning has yielded rates of transit ridership increasing substantially faster than rates of either overall population growth or vehicle miles traveled per capita. In fact, the most recent figures from the Federal Highway Administration suggest that this region is actually registering a slight decline in vehicle miles traveled per capita, exactly the opposite of the experience observed in other comparable US metropolitan areas (Metro (2003b)).

Participation Matters

Citizen involvement in planning is not a particularly dramatic notion these days. There are several reasons for seeking widespread participation in local planning efforts. Participation offers cover and legitimacy to decision makers and planners. However, it also contributes two essential elements that contribute to successful plans. First, widespread participation increases the range of ideas at the table. More eyes on the problem means more insight into its real nature and possible solutions (Hwang 1996). Second, involving a wide range of folks makes ownership of the results, the plan, widespread, particularly over the years and decades that it takes to act on plans.

Planning is an important part of the kind of community building needed to sustain values and visions over the long haul. Participation made Metro's Region 2040 planning process possible. In the Portland way, a "Noah's Ark" of a committee was constituted to guide the process of developing the RUGGO's, with membership consisting of elected officials from throughout the region, environmentalists, home builders, business interests, and citizens.

At the end of the two-year RUGGO process, it was suburban elected officials who observed that although they and Metro had developed a great description of the pieces of a well-functioning region, there was no vision, no overall description of where we were heading. Hence, it was the participants themselves that called for the creation of the Region 2040 Growth Concept, not Metro. Metro's signature growth management planning effort would never have happened without the participation of a wide range of interests.

Leadership Matters

Unquestionably, Oregon and the Portland region have benefited from terrific leaders that have made critical contributions at different points in time. Former Governor Tom McCall and key legislative leaders from both parties made the Oregon statewide land use planning program a reality. Former Governor and Mayor of Portland Neil Goldschmidt has been widely credited with leading the revival of Portland's downtown, neighborhoods, and civic culture. Today the region continues to benefit from key leaders in critical positions, people able to accurately name the problem, see the connections between interests, and inspire people to be better than they are.

However, the role for leadership has changed substantially in the last 25 years. Leaders no longer control information like they used to. Neighborhood activists run ArcView on their home computers. Information is now everywhere. Furthermore, in this region, interests are divided finely enough that there are no longer single "lead" issues that dominate the public policy agenda, pull interests together, and focus their efforts.

Good Things Take Time

The region is a work in progress. It's not done and it never will be. Things weren't always like they are today. Lewis Mumford visited the Portland region in the 1930's at the behest of a group of leaders investigating the ways that electrification could be integrated into the Pacific Northwest. Mumford had this to say about Portland and Seattle:

> *"...neither Portland nor Seattle show, from the standpoint of planning, more than metropolitan ambitions that have over-reached themselves. The melancholy plan to increase Portland's population from 300,000 to three million succeeded in disordering and unfocussing its growth: but it did little to give it the benefit of modern city planning practice; meanwhile, the apparent financial prospects of these port cities undermined the base of the sounder development that could well have been taking place in other parts of the region, on strictly modern lines." (Mumford (1939))*

His solution: Build new towns in a landscape that literally took his breath away. Some 60 years later, downtown Portland is a national model, but downtown was being rebuilt in the depths of the recession of the 1980's because of initiatives put into motion 20 and 30 years before. Creating a great place is the legacy of a lifetime, not a matter of months, a single term of office, or the cutting of a single ribbon. Stewardship takes place over generations, not all at once.

You Can't Regulate Quality of Life into Existence

Planners have lots of tools for limiting damage. However, community quality of life is a collective achievement. The great parks of this region didn't result from exactions, but from bond measures and other collective action. Modeling work done for the Region 2040 Growth Concept found that even with the ideal arrangement of buildings with relation to the transportation system, transportation system goals would not be met without changes in trip making behavior.

That is, absent a culture of inhabitation to go along with the physical development of the landscape, plan objectives would not be realized and community values wouldn't be served. Achieving quality of life goals is a partnership between what is required and the choices made by individuals. Regulation alone is not sufficient. Putting choice into context remains a central challenge for planners.

Sustainability is Ultimately about People, not just Places

Cities exist as places for interaction and exchange. Economies grow and thrive because of what people are able to do with each other, and the ways in which information and knowledge gets combined and recombined into forms of value to local communities and the world. Though this story here is told in terms of the "stuff" of urbanism --- land, transportation infrastructure, institutions, etc. --- it is ultimately a story of what people do with each other, on behalf of both each other and the places they live. Though thinking about the future is a distinguishing human trait, the future is, nonetheless, fundamentally uncertain.

Coping with uncertainty is accomplished in a variety of ways, including via avenues that involve planning, participation, leadership, and investment. However, perhaps the best hedge against uncertainty and in favor of sustainability is the broad engagement of people with each other. Sustainability, then, is never assured, but by recognizing that many perspectives develop resilience, it might be approximated if we understand that it comes not from perfecting our planning tools, but from enhancing our capacity to listen to, learn from, and act with each other.

Notes

Parts of this chapter have appeared in Seltzer (2005).

References

Abbott, C. (2001) *Greater Portland: Urban Life and Landscape in the Pacific Northwest*. University of Pennsylvania Press

Abbott, C. and Abbott, M.P. (1991) *Historical Development of the Metropolitan Service District*. Metro

Abbott, C., Adler, S. and Abbott, M.P. (1997) *Planning a New West: The Columbia River Gorge National Scenic Area*. Oregon State University Press

Downs, A. (2002) "Have Housing Prices Risen Faster in Portland than Elsewhere?", *Housing Policy Debate*, 13(1)

Hwang, S.W. (1996) "The Implications of the Nonlinear Paradigm for Integrated Environmental Design and Planning", *Journal of Planning Literature*, 11(2)

Knaap, G., Hopkins, L.D. and Ding, C. (1999) *Do Plans Matter? Effects of Light Rail Plans on Land Values in Station Areas* (Working Paper WP99GK1, Lincoln Institute of Land Policy)

Knaap, G., Bolen, R. and Seltzer, E. (2003) *Metro's Regional Land Information System: The Virtual Key to Portland's Growth Management Success* (Working Paper WP03GK1, Lincoln Institute of Land Policy)

Leccese, M. and McCormick, K. (2000) *Charter of the New Urbanism*, McGraw-Hill

Metro (1991) *Regional Urban Growth Goals and Objectives*, Metro

Metro (1995) *Report of the Future Vision Commission*, http://www.metro-region.org/article.cfm?ArticleID=433 (Accessed on 19 February 2004)
Metro (1997) *Regional Framework Plan*, http://www.metro-region.org/article.cfm?ArticleID=432 (Accessed on 19 February 2004)
Metro (2000a) *Regional Affordable Housing Strategy*, http://www.metro-region.org/article.cfm?articleid=417 (Accessed on 19 February 2004)
Metro (2000b) *The Nature of 2040: The region's 50-year strategy for managing growth*, http://www.metro-region.org/article.cfm?ArticleID=422 (Accessed on 22 February 2004)
Metro (2001) *Land Acquisition Report to Citizens*, http://www.metro-region.org/article.cfm?articleid=469 (Accessed on 19 February 2004)
Metro (2002) *2002 UGB Decision and Map*, http://www.metro-region.org/article.cfm?articleid=2504 (Accessed 19 February 2004)
Metro (2003a) *Metro Charter*, http://www.metro-region.org/article.cfm?ArticleID=3724 (Accessed 19 February 2004)
Metro (2003b) *The Portland Region: How Are We Doing?*, http://www.metro-region.org/article.cfm?ArticleID=3795 (Accessed on 19 February 2004)
Mumford, L. (1939) *Regional Planning in the Pacific Northwest: A Memorandum*, Northwest Regional Council
Nelson, A.C., Pendall, R., Dawkins, C.J., and Knaap, G.J. (2002) *The Link between Growth Management and Housing Affordability: The Academic Evidence* (Center on Urban and Metropolitan Policy, The Brookings Institution)
Oregon Department of Land Conservation and Development (2001) *Oregon Statewide Land Use Planning Goals and Guidelines*, http://www.lcd.state.or.us/goalhtml/goals.html (Accessed on 19 February 2004)
Ozawa, C. (ed) (2005) *The Portland Edge: Challenges and Successes in Growing Communities*, Island Press
Phillips, J. and Goodstein, E. (2000) "Growth Management and Housing Prices: The Case of Portland, Oregon", *Contemporary Economic Policy*, 18(3)
Seltzer, E. (2005) "It's Not and Experiment: Regional Planning at Metro, 1990 to the Present", in Ozawa, C. (ed) *The Portland Edge: Challenges and Successes in Growing Communities*, Island Press

15. Sustainable Urban Regeneration in Phoenix, Arizona: Implications for Multi-dimensional Governance

Carlos Balsas

15.1 Introduction

Human activity on earth and the increasing gregarious movement into cities are some of the most fascinating enterprises of mankind. This impact is difficult to measure, but Ehrlich and Holdren (1971) and Redman and Jones (2005) have defined it as a composite measure of population, affluence, technology and governance. Central to this equation is the long-term dimension of population impact and our ability to govern ourselves in order to minimize our impact on the planet. The sustainable development construct addresses intergenerational equity but also provides a framework to address city living and its regulatory milieu. Urban sustainability is now a major policy goal in many planning instruments and processes. Even though sustainable development means different things to different people, there is a consensus that sustainability is needed in order to achieve and maintain higher levels of economic prosperity, environmental quality and social responsiveness (Newman and Kenworthy (1999)); (Berke (2002)); (Gunder (2006)).

Central to the implementation of sustainability and urban management plans and projects is the concept of local and regional governance. Governance is the collaborative process of developing and implementing public policies. It involves not only issues of administrative power, public participation, but also the allocation of resources and the evaluation of specific plans and actions (Davies (2002)); (Hall (2004)); (Hall and Melnick (2004)). Sustainability and governance are two new paradigms that need to be integrated into institutional, professional and individual procedures and routines (Keil (2006)); (Kidokoro (2006)). Recently there has been a growing interest in governance processes to plan, deliver and implement sustainable development projects (OECD (2001)); (Moleketi (2003)); (McLennan and Ngoma (2004)). Murphy (2000) wrote that to achieve more sustainable cities, we need a system of strong urban governance based upon the principles of subsidiarity in a partnership model. Ayre and Callway (2005) edited a proceedings volume on the macro-implications of sustainability for governance. Fraser-Moleketi (2003) organized a conference in India about quality governance for sustainable growth and development. Several other pieces have been published about governance and sustainability in Europe (Astleithner and Hamedinger (2003)); (Lafferty (2005)).

In the US planning circles, there has been some research on governance issues both at the neighborhood (Arefi (2003)); (Portney (2005)) and the metropolitan levels (Collin, Leveillee and Poitras (2002)); (Hamilton, Miller and Paytas (2004)). But little efforts have been made to connect it with the sustainability literature. The same is true in Arizona. Here, there have been several pieces written on urban governance but also without focusing on sustainable development. Worth mentioning is the work of Moya (1998), Berman (1998), Hall (2002, 2004), and Hall and Melnick (2004).

The first thought somebody who has never been to Phoenix might have about it, is that the spread-out development of Phoenix is everything but sustainable. However, in this paper I argue that over the last two decades, the Phoenix metropolitan area has seen several attempts at influencing and implementing a more sustainable development agenda. On this agenda we find many urban regeneration projects, including housing infill projects promoted by the city of Phoenix; downtown revitalization projects in Phoenix and Tempe; waterfront regeneration projects in Tempe; multi-jurisdictional arrangements to build a new light rail system connecting three municipalities in the valley; the implementation of transit oriented developments (TOD); and major institutional commitments by Arizona State University, the largest university in the US, at teaching, researching, fostering and implementing the tenets of sustainable development on and off campus, among many others.

In this paper, I review recent urbanization trends in the Phoenix metropolitan area and discuss major urban regeneration projects and how they seem to be shaping new planning and governance practices in the valley. I argue that these projects and their underlying enabling governance processes constitute visible commitments and measurable indications to changing the perception and actual sustainability footprint impact of urban development in the Valley. To my knowledge this is the first attempt at linking the two discourses in Arizona.

This paper is organized into five main parts. Part one introduces and characterizes urban growth in Arizona. Part two discusses the main benefits and problems associated with rapid urban development. Part three presents and reviews a series of sustainable urban regeneration projects. Part four uses a vertical and a horizontal governance framework to analyze governance issues in the metropolitan area. This part ends with the 2003 Regional Transportation Plan as an example of a more sustainable metropolitan governance practice. Finally, part five presents some concluding remarks and recommendations.

15.2 The Phoenix Metropolitan Area: A Truly City-Region

Phoenix is located in the Sonoran desert in the southwest part of the United States. The capital city, Phoenix, is the fifth largest city in the country with 1.4 million people in 1,760 square miles. The Phoenix metropolitan area has 3.5 million inhabitants. But Arizona is now the nation's fastest growing state population-wise. The state grew 3.6 percent during the year ended July 1, 2006. Growth and Phoenix have been synonymous words for many years now. The metropolitan population has grown more than 45 percent in the last decade. The metropolitan area is 9,995 square miles, an area almost twice the size of Los Angeles County. The urbanized area of Phoenix metropolitan area comprises twenty-five cities and towns, three Indian communities and Maricopa County. Traditionally, the economy of the state has been characterized by the 5 C's: copper, cattle, cotton, citrus, and climate. More recently two other C's have been added to this list: construction and computers. Phoenix is an impressive metropolis only made possible by a very elaborate irrigation and water delivery system (Burns and Kenney (2005)), the stamina of its residents and business owners, the sharp rise of its population and the vision and leadership of its entrepreneurs.

The history of the settlement starts with the Hohokam Indians between 700 AD and 1100 AD. The Hohokam people were very successful with their farming and developed more than 500 miles of irrigation canals. At Pueblo Grande, it is estimated that the canal system irrigated at least 10,000 acres of farmland. The reason why this tribe extinguished is not well known, but popular belief argues that it outgrew the available resources in the area. Geographer Pat Gober (2006) tells us that the Anglo-European farmers arrived in the mid 1800s, followed by the Mormons. The mild and sunny weather in the region brought many health seekers to the valley, while the beauty of the desert magnified by the Grand Canyon brought many tourists as well. But Phoenix developed mainly during and after World War II. Military personnel migrated here due to the availability of jobs in the state's military bases. Many ended up staying because of the pleasant life and relatively cheap living. These characteristics also attracted many retirees when the real estate industry discovered the state in the 1960s with the construction of retirement communities, like Sun City, a Del Webb development (Luckingham (1989)). The region started attracting many partial-year residents, well known as snowbirds, who wanted to escape the winter rigors from the Midwest and New England states. The most recent wave of migrants is constituted mainly by Hispanic labor force to work in the booming construction and tourism industry.

In terms of urban development, Phoenix was relatively compact until World War II. Downtown aggregated most of the urban function in the city. Like many other cities, Phoenix developed a streetcar system. Central Avenue evolved over the years as the 'spine' of the metropolis. The first signs of decentralization occurred when the financial activities started moving north on Central Avenue. The appearance of automobiles and shopping malls also influenced the decline of the central city. Flat desert land allowed limitless urban developments and imaginations, including Frank Lloyd Wright's Broadacre utopian city. However, the growth of the suburbs, the continuous land annexations and the boom of retirement communities weakened downtown Phoenix's urban supremacy. In this context, the Phoenix airport, Sky Harbor, has played a pivotal role in urban development over the years. Its expansion in the 1960s and 1970s led to the relocation of many families to a booming suburb on the west side of the city: Maryvale (Yabes and Pijawka (2004)); (Collins (2005)). The damming of the Rio Salado in Tempe in the 1990s allowed the reenforcement of a centrality in the east valley. The growth of the metropolitan area during the last 50 years has been quite impressive; from little more than 100,000 people in 1950 to 1.4 million in 2000 (Heim (2001)).

The projects for the future of Phoenix are not less ambitious. The region is expected to grow to more than five million people by 2030. It is expected that most of the population growth will happen on the fringes of the metropolitan area, mainly in Buckeye and Queen Creek (MAG (2005)). In addition, research work by Robert Lang on megapolitan development predicts that Phoenix and Tucson will eventually be connected and constitute one big megapolitan corridor named the Arizona Sun Corridor. This is expected to happen around 2050 and will ultimately comprise more than seven million people (Lang (2006)).

Of course this fast-paced development brings not only many opportunities but also a lot of urban problems. The opportunities mainly allow citizens to maximize their individual pursuits, no matter what they are. Among the many opportunities we find wealth creation through jobs in all sectors of the economy, and particularly in those sectors associated with the construction and tourism industries. However, it is well known that "when we pursue what we want as an individual, we don't necessarily get what we want as a community" (Lincoln Institute of Land Policy (2003)). Among the main problems associated with fast urban growth in Arizona we observe the following:

- Downtown and urban decline
- Loss of desert, agricultural land, open space, and fauna and flora
- Air pollution – brown cloud
- Noise pollution
- Traffic congestion
- Urban heat island
- Lack of affordable housing
- Socially unjust developments and practices

These problems are not exclusive to Phoenix, but the fast pace of development has definitely exacerbated them here in the valley. Phoenix is very often pointed out as an example of a sprawly metropolis. Phoenix ranks above Los Angeles and below Philadelphia in the sprawlometer tool developed by the organization Smart Growth America.

Land development has traditionally happened in the outskirts where land is cheaper and where there are fewer constraints on development (Heim (2001)); (Gober and Burns (2002)); (Gammage (2003)). The problem with leapfrog-type development is that land in consolidated urban areas stays vacant, eventually gets abandoned and associated urban infrastructure starts decaying (Pijawka, Blair and Guhathakurta, et al. (1998)).

According to the Lincoln Institute of Land Policy (2003), some key facts associated with fast urban growth are:

- On average, 3,300 new homes are built every month in Maricopa County
- 150 lane-miles of new road are built each year
- One acre of desert is lost per hour to development in Metropolitan Phoenix
- Since 1992, Phoenix has built an average of six new golf courses per year and now has over 165 golf courses, according to the Arizona Golf Association
- The desert around Phoenix is the second most diverse ecosystem in the world, and contains more reptile species than any other ecosystem, including the tropical rainforest
- 80% of Arizona's population lives in the Sonoran desert
- In the next 10 years, more than two million people will move to the Phoenix area

Fig. 15-1. Key factors associated with urban growth in Phoenix (Source: http://www.makingsenseofplace.org/)

> "The Sonoran Desert is the second most diverse ecosystem on Earth. We have more plant and animal species here than anywhere else on Earth, except the rainforest. . .
> So that's why we talk about the developers all the time, and give them such a bad time. Because once they develop this little desert, it's gone, and a whole ecosystem goes with it that can never be replaced and exists nowhere else."
> - Tom Tucker, Walk Softly Tours
>
> "The desert is the essential magic of Phoenix. And we have the most spectacular and beautiful desert in the world."
> - Grady Gammage, land use attorney

Fig. 15-2. Ecological value of the Sonoran Desert (Source: http://www.makingsenseofplace.org/)

In addition, suburban residents have to rely on automobiles for their long daily commutes, which ultimately increases energy use and causes major air pollution and congestion problems (Newman and Kenworthy (1999)).

Some have pointed out that the Sonoran desert is one of the most diverse ecosystems on earth. If there is something that gets irreplaceably destroyed by urban development in Phoenix, it is desert land (Grymm and Redman (2004)). So over the years, different governmental institutions have attempted to preserve their desert lands. Maricopa Association of Governments developed a Desert Spaces Plan in 1995. The city of Phoenix approved its Sonoran Preserve Master Plan in 1998 and the city of Scottsdale approved the McDowell Sonoran Preserve in 2001. A very critical piece of urban growth in Phoenix is the handling of State Trust Land. In Arizona, as in most western states, a very large percentage of land is publicly owned. Unlike most public land, state trust land is held by the state for a specific purpose: to make money for education (Gammage and Fink (2004)). Over the years this land has generated a lot of controversy. On one side there are those who see it as a commodity with exchange value, and on the other there are those who regard it as having intrinsic value which is worth preserving for future generations.

Of course, municipalities are aware of the negative consequences caused by fast urban growth. In a 1997 article, then assistant planning director for the city of Phoenix and his colleagues (Quay, Mathien and Richert (1997)) developed a strategic view of growth in Phoenix based on the following principles:

- Future employment growth within all employment centers will be closely linked to the characteristics and growth of the residential areas within the center's commute shed
- Future sales tax generation will be more tightly linked to providing retail opportunities close to residential areas

- The ability to financially provide infrastructure will be the major constraint for residential growth around the urban fringe
- The commute shed for some employment centers will eventually not include residential growth areas at the urban fringe so urban infill becomes key to employment growth of such centers

Urban growth in Phoenix does not respect administrative boundaries. The State of Arizona has enacted several regional growth management strategies over the last decade. The Arizona Preserve Initiative Program (API) was passed by the legislature in 1996. Its main goal was to encourage the preservation of state lands near urban areas as open space for future generations. The Growing Smarter legislative package was enacted in 1998. This legislation aimed at strengthening city and county planning by requiring that new general and comprehensive plans be re-adopted regularly and that all subsequent zoning actions conform to the plans, as well as to change the way state trust lands are managed by emphasizing open space preservation. This act was subsequently re-authorized by the Growing Smarter Plus legislation in 2000. These three strategies made significant changes in the way city and county governments plan and regulate their lands, citizens play a role in land use issues, state trust lands are managed, and open space may be acquired and preserved (Heffernon and Melnick (2001)).

More recently, the Growing Smarter Guiding Principles were developed in 2006. The Arizona Growing Smarter Oversight Council has been given

- Water: developing reliable information for all areas of the state and incorporating long-term available assured supplies in all future planning efforts.
- Transportation: ensuring efficient, effective transportation systems and links both locally and regionally simultaneous with future development.
- Housing: providing an array of workforce and affordable housing choices to assure that people can live in the same community where they work.
- Community Character: preserving a sense of place and distinctive identities
- Education: assuring quality schools, education and workforce development opportunities to prepare for global competition.
- Economic Development: pursuing stable, diverse economies statewide and preserving traditional economic sectors, such as agriculture, while moving into the global knowledge economy.
- Local Authority: providing cities and counties the powers and means to manage new development, especially giving counties an array of statutorily authorized powers to manage growth, including subdivisions and lot splits.
- Energy: making the necessary investments and providing incentives in reliable energy supplies and alternative energy sources.

Fig. 15-3. Growing Smarter Guiding Principles 2006 (Source: http://www.azcommerce.com/)

Placed on the November 2000 ballot by the Arizona Legislature, proposition 100 would have amended the state constitution by designating up to 3% of state trust land (approx. 270,000 acres) for permanent conservation. It would have also established guidelines for exchanges of trust lands and donations for school sites and extended agricultural and grazing leases beyond ten years, without auction. Proposition 100 had been struck down by the Arizona Superior Court in August, but the state's appeal was upheld by the Arizona Supreme Court. This proposition was defeated by a margin of 52% to 48%.

Fig. 15-4. Proposition 100 (Source: http://www.urbanfutures.org/)

The Citizen's Growth Management Initiative, drafted by the Sierra Club, appeared on the November 2000 ballot in Arizona and was defeated by Arizona voters by a margin of 70% to 30%. It would have required each municipality and county in the State with a minimum population of 2,500 persons to adopt a 10-year growth management plan. Its provisions would have:

- required cities and counties to establish urban growth boundaries
- required additional impact fees
- required plans and policies to be consistent with state/federal air and water quality standards
- placed additional protections on ground and surface water supplies
- established policies protecting environmental quality, neighborhoods, scenic vistas, open space, historic/archaeological sites and provide for affordable housing
- promoted multi-modal forms of transportation
- required citizens to vote on zoning changes and boundary extensions

Fig. 15-5. The Citizen's Growth Management Initiative (Source: http://www.urbanfutures.org/)

the responsibility of continuously assessing the successes and challenges of Arizona's planning and growth management legislation, the 1998 and 2000 Growing Smarter Acts. Grassroots processes developed a set of Guiding Principles to ensure that Arizona's growth maintains and enhances quality of life. Over 1,300 people from around the state participated in 18 Listening Sessions and a statewide telephone poll during the summer and fall of 2005. The Guiding Principles were approved by the Growing Smarter Oversight Council in August 2006.

Throughout the years, other urban growth strategies have not received positive support from the voters. Among those, we find proposition 100 regarding the state trust lands and proposition 202, known as the Citizen's Growth Management Initiative.

More recently, two other propositions have been rejected by the voters in the November 2006 elections. These were propositions 105 and 106 on state trust lands. The former was intended to ensure and increase the economic value of state trust land for the benefit of public schools and the other beneficiaries, while providing opportunities for conservation consistent with the mission of the state land trust. The later would create a new Board of Trustees appointed by the governor to plan and dispose of all state trust lands, and set aside 694,000 acres of land that could not be sold or developed.

15.3 Sustainable Phoenix: An Oxymoron?

Sustainability is an umbrella concept for economic development, environmental conservation and social responsiveness. At first Phoenix does not look very sustainable; however, upon close examination there seems to be a shift in the horizon. This shift might be the result of negative externalities outgrowing the positive, as well as the growing awareness of some potential constraints on urban growth, such as water and a slower-paced economy, and overall, the realization of quality-of-life nuisances.

More recently, sprawly environments do not seem to be very popular among the creative class, which likes to come together in bustling urban areas to mingle and exchange ideas. In terms of urban policy, research on urban revitalization shows that Phoenix is good at delivering big item developments, e.g. the big convention center, the planned united development in the outskirts, but that it is time to revamp revitalization practices and start playing the 'inside game' (Morisson Institute for Public Policy (2003)). In this sense, we argue that several urban regeneration projects seem to fit the goals of an emerging sustainability agenda, and these projects are:

- Neighborhood infill redevelopments in Phoenix
- Downtown revitalization projects in Phoenix and Tempe
- Waterfront regeneration projects in Tempe
- Multi-jurisdictional arrangements to build a new light rail system connecting three municipalities
- The implementation of transit oriented developments (TOD)
- Arizona State University's commitment to sustainability

The Phoenix infill housing started in 1995 when Phoenix authorized funds for single-family housing on vacant land in the central area. 3,175 new single-family homes have been built, about one-third of these are affordable for low- and moderate-income families with an estimated construction value in excess of $150 million. In addition, pilot programs are in place to address regulatory requirements, plan review times, and costs associated with city review (Heim (2001)); (City of Phoenix (2006)).

The revitalization of downtown Phoenix started in the 1970s with the construction of the Civic Plaza comprising the Convention Center and the Symphony Hall. This was followed by the implementation of the redevelopment area in 1979. The Arizona Center shopping center, as well as several museums, the ball park, and sports arena, all were built to capitalize on the central location and help revitalize the downtown area. These two latter ones were built with public support. Instrumental in this process has been the creation of a business improvement district called Copper Square. If the first objective of this new organization was to create a 'clean and safe' downtown, now it is getting more involved in the promotion of redevelopment opportunities. The convention center is also being renovated and the first phase was recently completed this year.

The most recent attempt at revitalizing the downtown area is the construction of a new campus of the Arizona State University in downtown Phoenix (ASU (2004)). When fully completed by 2015 the campus will serve an estimated 15,000 students and 2,000 staff and faculty. This development is accompanied by a new biomedical campus also in downtown Phoenix centered on the Translational Genomics Research Institute (TGen), the College of Medicine and the Bioscience high-school (Kearney (2006)). There are also several projects being developed for the Capitol mall district (Pursue a centennial dream (2005)); (Balsas (2006)). Finally, the city of Phoenix has embarked on a major project to revise the regulatory environment governing the planning of the downtown area. This is known as the Phoenix downtown urban form project.

Tempe has also experienced an urban revitalization face-lift in the last decade. The northern section of Mill Avenue has been narrowed down to two lanes and developed according to a new streetscape pattern. Due to its close proximity to ASU, this true main street comprises many restaurants, bars and a lively street scene (Davis (2004)). A very ambitious project also in the city of Tempe is the Tempe Town Lake that comprised the creation of two dams on the Salt River and the redevelopment of its banks.

The Phoenix light rail systems is a major transportation project involving the construction of 20 miles of light rail connecting three cities in the metropolitan area: Phoenix, Tempe and Mesa. It has been budgeted at $1 billion and the project is expected to be completed by the end of 2008. This light rail system was developed not as an alternative to the freeway system in the metro area, but as one more option to commuters in the central part of the metropolitan area. The merit of this transportation system is reflected in the redevelopment opportunities it brings with it (ULI (2001)). The station areas are seen as major opportunities for infill development and the three cities are in the process of developing Transit Oriented Development ordinances to facilitate and encourage redevelopment opportunities (City of Phoenix (2004)). At the moment, future expansions of the current system are being debated by different cities in the metropolitan area.

Finally, Arizona State University has shown a very positive commitment to sustainability issues. The genesis of this commitment goes back to the existence of a very accomplished research center, the Center for Environmental Studies (CES), now re-named Global Institute of Sustainability (GIOS), and the president's vision for ASU as a New American University. GIOS has had many National Science Foundation (NSF)-funded projects in the area of urban ecology and environmental studies. In fact, the Central Arizona-Phoenix Long-Term Ecological Research has produced very comprehensive research on urban heat island, fauna and flora and the urban ecology of Phoenix. The Greater Phoenix 2100 Atlas is a product of CES.

One of the main design principles of the New American University is community embeddedness. With that in mind, the university has created a new School of Sustainability, which will offer undergraduate and advanced degrees in sustainability science starting in 2007. Besides this localized attempt, there are many other university-wide initiatives, such as the Decision Theater, a multimillion-dollar state-of-the-art facility to facilitate the visualization of new developments and their multiple impacts, and the Office of Sustainability Initiatives, which just organized a major retreat on how to streamline sustainability throughout the curriculum and across departments in September 2006.

"There are more and more elected officials saying . . . not "no to growth," but no to the way we are growing, and yes to a look at what our other options are. And that means, I believe, more cooperation between the cities --- and we're seeing some --- and ultimately some kind of an over-arching planning process. We can't do that as 24 warring communities."

- Terry Goddard, Mayor, City of Phoenix, 1984-1990

"What affects my city affects all the cities surrounding. If we don't work as a region, we don't have a very good place to live."

- Joan Shafer, Mayor, City of Surprise

Fig. 15-6. The need for more regional cooperation in Phoenix (Source: http://www.makingsenseofplace.org/)

15.4 Governance in Phoenix, AZ

Governance is the collaborative process of developing and implementing public policies. It involves not only issues of administrative power and public participation, but also the allocation of resources and the evaluation of specific plans and actions.

It is well known that most complex challenges today have a regional scale and that traditional business and government structures and civic responses are not adequate. The urban challenges in the 21st century require multi-sectorial and multi-disciplinary approaches, and very few stakeholders know how to engage in this kind of regional civic leadership (Peirce, Johnson and Hall (1993)); (Brown and Chapman (2001)).

In the words of Hall and Melnick (2004), governance in Phoenix is "governance light and mainly reactive." Tremendous competition for growth over the years has fueled unrestrained development and land speculation, where the fragmented structure of local governments has led to weak regional planning capacity.

Drawing on some previous literature (Hamilton, Miller and Paytas (2004)), I suggest that governance for sustainability in Arizona requires better vertical and horizontal integration. In terms of vertical levels, governance should involve the following stakeholders: citizens, cities and towns, Maricopa County, Maricopa Association of Governments and the State of Arizona. In terms of horizontal governance, we can identify, among others, at least the following three dimensions: economic, environmental and social.

Citizens ought to be the ultimate recipients and supposedly the first initiators of effective governance processes. And because of this, they appear at the top of the governance pyramid. They have rights and

According to a report published by the Morrison Institute for Public Policy (2000), governance in Phoenix is influenced by the following factors:

- a 'frontier spirit' continues to shape the region
- cities and city-based localism predominate
- land use control is seen as a local rather than a state or a regional activity
- emphasis on the sales tax as a source of municipal revenues has encouraged intense competition for the location of economic generators
- Maricopa County and the Maricopa Association of Governments (MAG) lack strong authority
- the region's civic culture remains fractious
- the State of Arizona has provided inconsistent leadership in the metropolitan area

Fig. 15-7. Main Characteristics of Governance in Phoenix

responsibilities, they can attend public meetings, and they can get organized and/or run for office. Individual participation in common affairs is very important for civic and social capital to develop overtime. Hall (2001) mentions the work of Pierce to demonstrate that "there is an increase in civic capacity in proliferating neighborhood organizations and community development corporations, urban youth self-help ventures, and regional strategic planning efforts." But we know that a responsive government capable of providing and managing public services efficiently is a guaranty to many citizens that they can settle in a certain municipality and get involved in public affairs. Because ultimately, "people will vote with their feet" and they will move to the municipality that provides the best bundle of goods and services (Tiebout (1956)), if the above level of services does not happen. So, it is in the government best interest to share and involve citizens in the running of public affairs.

Cities in the valley are characterized by a strong council-manager system, where the city managers run their cities, the council members represent their districts and mayors are looked at for leadership. Planning occurs mainly at the village level and projects are approved by the planning commission. The most typical planning instruments and players in Phoenix are: the general plan, the planning commission, the board of adjustments, planning village committees, design review standards committee, city codes and ordinances (subdivision, zoning, etc.), development impact fees, area and neighborhood plans, long range growth strategies and freeway mitigation plans. McCabe (2006) reveals that enclaves of private services coordinated by Home Owners Associations (HOA) also influence the city's planning and provision of public services (see also Kirby, A. et al. (2006)).

With the following three recent honorable exceptions, cities rarely collaborate among themselves: 1) the mayor of Phoenix's announcement of Washington Street as an opportunity corridor (Kress (2005)), 2) the mayor of Phoenix's request to stop giving tax breaks to big-box retailers (Gordon to cities: Tax lures must end, 2005), and finally 3) the Arizona centennial celebration in 2012 is likely to bring to the foreground some collaborative processes around centennial projects to honor the history and accomplishments in the state.

Most of the urban growth in the Phoenix metropolitan area has occurred within Maricopa County. Maricopa County is a player in the metropolitan governance equation because it provides planning and development services to constituents of unincorporated Maricopa County, so that they can develop responsibly.

Regional planning in Arizona mirrors the process found in the other 49 states. Regional planning is provided through regional councils of governments. Maricopa Association of Governments (MAG) is the Metropolitan Planning Organization in Phoenix. In order to qualify for federal assistance on regional projects, Arizona councils of government were organized into formal bodies. Arizona Regional Councils have grown into pro-active, multi-functional, full-services regional organizations prepared to enable regional communities, their leaders, and citizens to compete and thrive locally. The federal, state and local agencies and tribal governments also conduct regional planning and coordination activities (e.g. Active Management Areas of the Arizona Department of Water Resources, and Growing Smarter Commission (2000)).

MAG is a council of governments that serves as the regional agency for the metropolitan Phoenix area. MAG is a forum for discussion and study of regional problems in the areas of areas of transportation, air quality, environment, regional development and social services. It ensures pooling of

Transportation • 20-year multi-modal Long Range Transportation Plan • 5-year Trans. Improvement Program • Inter-modal Management Systems • MAG Freeway Program • Streets • Grand Avenue EIS • Pedestrians, bicycles • Elderly & persons with disabilities	Air quality • Clean Cities Alternative Fuel Plan and Program	Water quality and waste water • Prepares 208 Plan
Land use • Enhanced Notification Process • Urban Growth Studies/Plan • Open Space Plan • Coordinate and consult with federal and state agencies regarding military bases, natural resources, land management	Social • Human services to children, adults, elderly and persons with physical and developmental disabilities • Juvenile justice and delinquency • 911 coordination • Library services • Youth issues • Domestic violence • Census	Economic • Population estimates and projections • Building codes • Specifications and details for public works • Construction

Fig. 15-8. Types of planning currently done by MAG (Source: Growing Smarter Commission (2000))

resources; plans for the solution of regional problems requiring multi-city, town and county cooperation; facilitates agreements for the adoption of common policies; and finally, prepares for future growth and development of the region.

The MAG membership consists of the 25 incorporated cities and towns within Maricopa County and the contiguous urbanized area, the Gila River Indian Community, the Salt River Pima Maricopa Indian Community, Fort McDowell Yavapai Nation, Maricopa County, the Arizona Department of Transportation (ADOT) and the Citizens Transportation Oversight Committee (CTOC). ADOT and CTOC serve as ex-officio members for transportation-related issues (MAG (2005)).

MAG developed a statement of regional values in 1994 and was quite instrumental in bringing together tens of stakeholders to support the development of a regional vision in 2000: Valley Vision 2025. Aware of the need to improve its regional governance practices, MAG created a task force in 2001 to study issues of regional governance and to pursue ways to further engage MAG in the community.

According to MAG (2002), the outcomes endorsed by the MAG Regional Council included the following nine conceptual actions:

1. That MAG delegate more responsibility and authority to an expanded Executive Committee, comprised of members from the MAG Regional Council.
2. A super majority would be required at the Executive Committee to overrule a policy committee's recommendation. In the event there is no super majority, but a simple majority, the decision would be forwarded to the full Regional Council.
3. That MAG expand the business and other stakeholder representation on all of the MAG technical and advisory committees.
4. That MAG hold Town Hall meetings, with a specific identified invitation list, that will advise the Regional Council on goals.
5. That the Regional Council consider whether it wants to extend voting participation to local governments in the contiguous urbanized area.
6. That there be a compilation and report of the impact of proposed significant developments on the regional facilities for transportation, open space, wastewater and water, prior to the jurisdiction taking action on the development.
7. That data compiled on developments of significance would be integrated into the MAG Transportation Improvement Program and Long Range Transportation Plan, and the MAG 208 approval processes, where there is an action going forward to the Regional Council.
8. That MAG integrate local and regional transit planning into the regional transportation system planning effort in cooperation with Valley Metro.
9. That after these recommendations are accepted, council members look at re-branding MAG.

Fig. 15-9. Re-conceptualizing governance in Phoenix

MAG planning efforts have been quite substantial in the transportation arena. The state created a Regional Public Transportation Authority in 1985 and the county voters approved a one-half-cent sales tax for freeway and transit service through 2005. Different cities approved legislation that would dedicate small fractions of their sales taxes to improving the regional transit system:

- Tempe: 0.5 cent sales tax in 1996
- Mesa: quality-of-life 0.5 cent sales tax in 1998
- Phoenix: 0.4% sales tax in 2000
- Glendale: 0.5 cent sales tax in 2001

MAG was quite instrumental in developing the Regional Transportation Plan, which was finally approved in 2003. This plan was the evidence needed to bring Proposition 400 to the ballot in 2004. The county voters approved an integrated roads and transit funding package that will provide $9 billion in regional transportation revenue over the next 20 years (Regional Transportation Commission (2006)).

The Regional Transportation Plan is significant for the following reasons: 1) it is the most comprehensive transportation planning effort in 40 years, 2) it provides a blueprint for transportation investments for the next 20 years, 3) it serves as a foundation for half-cent sales tax extension, and 4) it was developed by the Transportation Policy Committee – a new way of doing business for MAG. The plan was the outcome of hundreds of public meetings, workshops, and hearings to facilitate public involvement and input in something that will affect them greatly in the future. The basic premise of the plan is that growth and increase travel will require improvements in all modes: freeways, highways and arterials, regional bus system, and bus rapid transit and light rail.

Horizontal governance dimensions include among many others, economic, environmental and social mobilizations to make government a more transparent and effective activity. The valley is has many different types of non-governmental organizations that shape the way urban growth and public affairs are conducted on a daily basis. Worth mentioning because of their involvement in many public causes are the Greater Phoenix Economic Council, the Valley Forward and Chicanos por la Causa.

15.5 Conclusions and Recommendations

Phoenix's fast urban development does not seem very sustainable in the long term. I argue that there are several sustainable urban regeneration projects that can lead to a more sustainable path. Their underlying enabling governance processes constitute commitments and measurable indications to changing the perception and actual sustainability footprint impact of urban development in the valley.

However, Berman (1998) as argued that:

> *"There is no entity to prevent ruinous competition among neighboring cities, nor is there any effective way to ensure that cities consider the effects of their developmental decisions on neighboring cities. The state is involved in growth management through its control of State Land Trusts (i.e. annexation should be made easier and incorporation more difficult). The future of growth management is largely in the hands of the state government: 1) it could impose rigorously enforced and detailed planning mandates under which all local plans must fold into regional plans and the regional plans into a state plan. 2) alternatively, it could require regional planning which recognizes that Arizona is relatively decentralized and locally diverse. A 'one size fits all approach' is not appropriate, and state mandates need to be accompanied by ample technical and financial assistance to jurisdictions that need help."*

Among the most important recommendations one can give are:

- The need for a land use regulatory review for developments having significant regional impacts.
- That social responsiveness should be brought more fully to the sustainability agenda.
- That there is need for more regional binding procedures.
- That there is need for more effective arenas for discussion.
- That more citizen participation and involvement (bottom-up) will strengthen governance practices.
- That more state leadership (top-down) can enable more effective governance processes.

References

Arefi, M. (2003) "Revisiting the Los Angeles Neighborhood Initiative (LANI)", *Journal of Planning Education and Research,* 22

Arizona Growing Smarter Oversight Council (2006) *Growing Smarter Guiding Principles 2006,* http://www.azcommerce.com/CommAsst/GrowSmart/Guiding+Principles.htm (Accessed on December 20, 2006)

Astleithner, F. and Hamedinger, A. (2003) "Urban sustainability as a new form of governance: obstacles and potentials in the case of Vienna", *Innovation,* 16(1)

ASU (2004) *Downtown Phoenix campus*, http://www.asu.edu/ cdp/dtn.html (Accessed on December 20, 2006)
Ayre, G. and Callway, R. (eds) (2005) *Governance for Sustainable Development, a Foundation for the Future.* Earthscan
Balsas, C. (2006) "The Phoenix Capitol Mall studios as examples of community embeddedness", *Open House International*, 31(3)
Berke, P. (2002) "Does sustainable development offer a new direction for planning? Challenges for the twenty-first century", *Journal of Planning Literature*, 17(1)
Berman, D. (1998) "The growth management challenge in Arizona", in Melnick, R. (ed) *Growth in Arizona, The Machine in the Garden.* Morrison Institute for Public Policy
Brown, B. and Chapman, J. (2001) *Pieces of Power: Governance in Arizona.* Arizona Town Hall
Burns, E. and Kenney, E. (2005) "Building and Maintaining Urban Water Infrastructure: Phoenix, Arizona, 1950-2003", *Yearbook of the Association of Pacific Coast Geographers*, 67
City of Phoenix (2004) *Transit Oriented Development TOD Guidelines – Phoenix*, http://www.valleymetro.org/METRO_light_rail/ (Accessed on December 20, 2006)
City of Phoenix (2006) *Phoenix Infill Housing Program.* http://phoenix.gov/ BUSINESS/infilpgm.html (Accessed on December 20, 2006)
Collin, J., Leveillee, J. and Poitras, C. (2002) "New challenges and old solutions: Metropolitan reorganization in Canadian and US city-regions", *Journal of Urban Affairs*, 24(3)
Collins, W. (2005) *The Emerging Metropolis: Phoenix, 1944-1973.* Arizona States Park Board
Davies, J. (2002) "The governance of urban regeneration: A critique of the "governing without government" thesis", *Public Administration*, 80(2)
Davis, L. (2004) "Finding room for history in the desert: Can Tempe afford-or afford not to-keep its oldest house?", *The Next American City,* February, http://www.americancity.org/ (Accessed on December 20, 2006)
Ehrlich, P. and Holdren, J. (1971) "Impact of population growth", *Science*, 171
Fraser-Moleketi, G. (2003) "Quality governance for sustainable growth and development", *International Review of Administrative Sciences*, 69(4)
Gammage, G. (2003) *Phoenix in Perspective – Reflections on Developing the Desert* (2nd Edition). Herberger Center for Design Excellence and Arizona State University
Gammage, G. and Fink, J. (2004) *The Phoenix Experiment* (Paper delivered at the conference "On the Edge: Metropolitan Growth and Western Environments" Stanford University, April 15-16)
Gober, P. (2006) *Metropolitan Phoenix: Place Making and Community Building in the Desert.* University of Pennsylvania Press
Gober, P. and Burns, E. (2002) "The size and shape of Phoenix's urban fringe", *Journal of Planning Education and Research*, 21
Gordon to cities: Tax lures must end. *Arizona Republic*, May 19, 2005.

Growing Smarter Commission (2000) *Regional Planning Issues Technical Advisory Subcommittee Report and Recommendations*, http://www.azplanning.org/commission.html (Accessed on December 20, 2006)

Grymm, N. and Redman, C. (2004) "Approaches to the study of urban ecosystems: The case of central Arizona – Phoenix", *Urban Ecosystems*, 7

Gunder, M. (2006) "Sustainability, planning's saving grace or road to perdition?", *Journal of Planning Education and Research*, 26

Hall, J. (2002) "Reconsidering the connection between capacity and governance", *Public Organization Review: A Global Journal*, 2

Hall, J. (2004) "Who will govern American metropolitan regions?", in Phares, D. (ed) *Metropolitan Governance Without Metropolitan Government?* Ashgate

Hall, J. and Melnick, R. (2004) "Regional roles, relationships and the future of the Phoenix metropolitan area", in Phares, D. (ed) *Metropolitan governance without metropolitan government?* Ashgate

Hamilton, D., Miller, D. and Paytas, J. (2004) "Exploring the horizontal and vertical dimensions of the governing of metropolitan regions", *Urban Affairs Review*, 40(2)

Heffernon, R. and Melnick, R. (2001) *Growth management and open space protection in Arizona: Current tools and progress* (Issues in Brief on behalf of Greater Phoenix Leadership) Morrison Institute for Public Policy

Heim, C. (2001) "Leapfrogging, urban sprawl and growth management: Phoenix 1950-2000", *American Journal of Economics and Sociology*, 60(1)

Kearney, B. (2006) *Copper square and downtown Phoenix: The next great American downtown* (Presentation made to the ULI Arizona District Council. August 30)

Keil, A. (2006) "New urban governance processes on the level of neighborhoods", *European Planning Studies*, 14(3)

Kidokoro, T. (2006) "An introduction to sustainable urban regeneration and local governance in regional cities" (Paper presented at the International workshop on Sustainable Urban Regeneration and Local Governance in Regional Cities. January 17-18)

Kirby, A. et al. (2006) "Examining the significance of housing enclaves in the metropolitan United States of America", *Housing, Theory and Society*, 23(1)

Kress, A. (2005) "Phoenix mayor calls for opportunity corridor" *Phoenix Business Journal*, March 18

Lafferty, W. (ed) (2004) *Governance for Sustainable Development, a challenge of adapting form to function.* Edward Elgar

Lang, R. (2006) *The sun corridor: Arizona's emerging megapolitan area* (Presentation delivered at the Arizona Town Hall, Prescott. April 9, http://www.mi.vt.edu/ (Accessed on December 20, 2006))

Lincoln Institute of Land Policy (2003) *Phoenix: The Urban Desert, a documentary film.* Lincoln Institute of Land Policy

Luckingham, B. (1989) *Phoenix: The History of a Southwestern Metropolis.* The University of Arizona Press

MAG (2000) *Valley Vision 2025 – Vision Report.* Maricopa Association of Governments
MAG (2002) "MAG Regional Council endorses governance concepts", *MAGA-Zine*, February
MAG (2003) *Regional Transportation Plan*, http://www.mag. maricopa.gov/ (Accessed on December 20, 2006)
MAG (2005) *Regional Report, A Resource for Policy Makers in the Maricopa Region.* Phoenix Maricopa Association of Governments
Maricopa County (2002) *Maricopa County Comprehensive Plan - MC2020*, http://www.maricopa.gov/planning/compln/ (Accessed on December 20, 2006)
McCabe, B. (2006) "Privatizing urban services through homeowners associations: the potential and practice in Phoenix", *International Journal of Public Administration*, 29(10-11)
McLennan, A. and Ngoma, W. (2004) "Quality governance for sustainable development?", *Progress in Development Studies*, 4(4)
Morrison Institute for Public Policy (2000) *Hits and Misses: Fast Growth in Metropolitan Phoenix.* Morrison Institute for Public Policy
Moya, S. (1998) "Growth management in Maricopa County, 1988-1992: An application of Kingdom's agenda setting model", *The Social Science Journal*, 35(4)
Murphy, P. (2000) "Urban governance for more sustainable cities", *European Environment*, 10
Newman, P. and Kenworthy, J. (1999) *Sustainability and cities.* Island Press
OECD (2001) *Cities for Citizens: improving metropolitan governance.* OECD
Peirce, N. Johnson, C. and Hall, J. (1993) *Citistates: how urban America can prosper in a competitive world.* Seven Locks Press
Pijawka, D., Blair, J., Guhathakurta, S. et al. (1998) "Environmental equity in central cities: Socioeconomic dimensions and planning strategies", *Journal of Planning Education and Research*, 18(2)
Portney, K. (2005) "Civic engagement and sustainable cities in the United States", *Public Administration Review*, 65(5)
Pursue a centennial dream (2005) *Arizona Republic*, December 8
Quay, R. et al. (2003) *GP2100 Greater Phoenix Regional Atlas.* ASU
Quay, R., Mathien, J. and Richert, D. (1997) *Phoenix's Strategic View of Growth,* http://www.design.asu.edu/apa/proceedings97/quay2.html (Accessed on December 20, 2006)
Redman, C. and Jones, N. (2005) "The environmental, social and health dimensions of urban expansion", *Population and Environment*, 26(6)
Regional Transportation Commission (2006) *Examples of regional governance,* http://www.psrtc.wa.gov/ (Accessed on December 20, 2006)
Tiebout, C. (1956) "A pure theory of local expenditures", *The Journal of Political Economy*, 64(5)
ULI (2001) *Light-rail transit in Phoenix, Arizona, economic development along the planned light-rail line,* http://www.valleymetro.org/ METRO_light_rail/ (Accessed on December 20, 2006)

Yabes, R. and Pijawka, D. (2004) *Public participation in achieving sustainability in central city neighborhoods* (Paper presented at the IUPEA 2004 conference in Louisville)

Part VI

CONCLUSION

16. Towards Sustainable Regeneration of City Regions

Tetsuo Kidokoro, Noboru Harata, Leksono Probo Subanu, Johann Jessen, Alain Motte and Ethan Paul Seltzer

16.1 Urbanization and Spatial Development Strategy

16.1.1 Sustainable Urban Regeneration

As the competition among city regions becomes increasingly harsh in the age of globalization, regional cities have increasingly be requested to play key roles as driving forces of city regions. Sustainable urban regeneration is understood as the regeneration of the attractiveness of cities in a sustainable manner in response to an ever-changing external world. How can this conceptual meaning of sustainable urban regeneration be interpreted in spatial terms? Fig. 16-1 illustrates the relationships between urbanization and the directions of urban spatial development. In the age of urbanization and motorization, selective redevelopment of city center areas and suburban development are facilitated at the same time and sprawl type of spatial development occurs (suburbanization stage). Generally, cities in the developing world are now at this stage of urban development. On the other hand, in the cities where urbanization has already reached a matured stage as observed in most developed countries, investment in extended urban areas becomes a mainstream trend as evidenced in the emergence of edge cities, and investment in old city center areas decreases significantly, in particular, in the old industrial areas. This stage of urban development can be called as the exsurbanization stage.

Urban regeneration strategy is the policy for rearrangement of spatial development patterns. This type of strategy is currently adopted in many cities in the developed world where cities are generally at the exurbanization stage. The urban regeneration strategy aims at attracting private

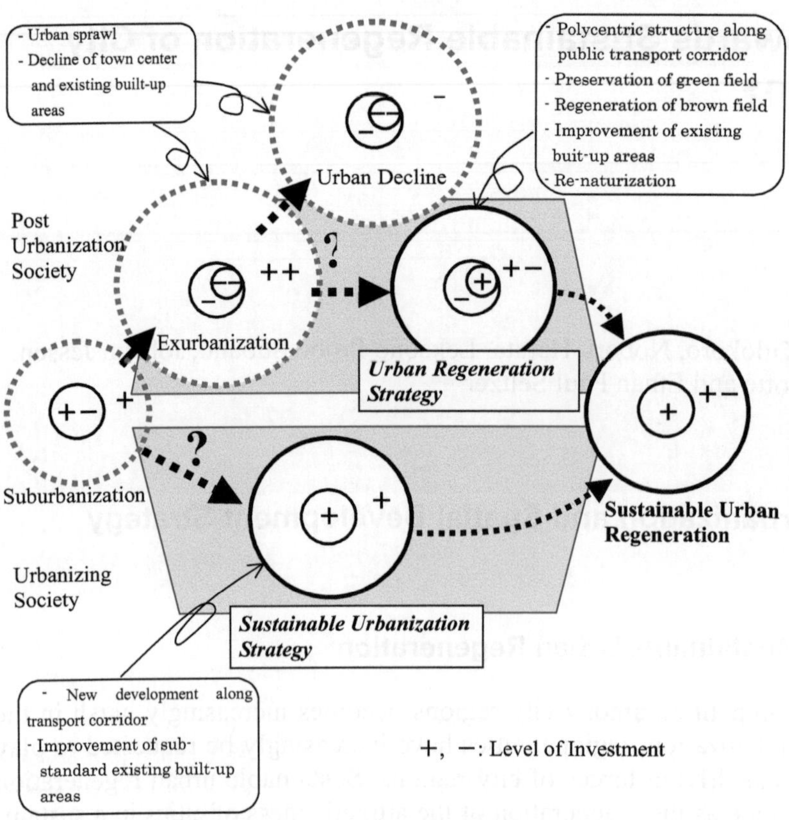

Fig. 16-1. Urbanization and spatial development strategies

investment in the city center areas, particularly in brownfield areas through urban regeneration projects which are jointly financed by the public and private sectors. At the same time, it attempts to limit the urban development in green areas and to guide development to public transport corridors in suburban areas in order to pursue social, environmental and economic sustainability. Among the case cities, it seems that Copenhagen, Stuttgart and Portland are successful examples of urban regeneration strategy. In these cities, the expansion of private investment in the city center areas has been triggered by urban regeneration projects and a land-use regulatory framework has been established to restrict urban development in green areas, though the level of restrictions varies. Judging from the experiences in these cities, establishing the institutional framework at the regional scale will be a key to sustainable urban regeneration.

On the other hand, in urbanizing cities of the developing world, as the sustainable urbanization strategy, it is important to develop efficient public transport systems and to guide urban development along public transport corridors both in the city center areas and rapidly urbanizing areas. Among the case cities, Dalian is a successful case in this regard.

The important factors in conducting urban regeneration strategy and sustainable urbanization strategy are examined below.

16.1.2 Sustainability as an Overall Objective

The overall objective of sustainability has been attached to urban development since the early nineties in the course of the debate on the outcome of the UN conference in Rio de Janeiro. Urban regeneration and development has to find a balance between economic, social, ecological and cultural objectives. This requires an integrated approach. Urban development strategies with different goals such as increasing regional economic competitiveness, preserving a city's quality of life and protecting the environment and saving energy have increasingly become interconnected.

16.1.3 Urban Regeneration and Development Projects Based on Local Initiative

In urban regeneration strategy, large-scale symbolic projects, such as the construction of LRT, waterfront development, major international events, international hub facilities such as international airports and high-speed rail etc., are implemented as key urban regeneration projects. In the case of the sustainable urbanization strategy, large-scale infrastructure projects, such as rail mass transit projects, are to be implemented as key urban development projects. National or federal governments (plus EU in the case of EU countries) have played major roles in funding. It is noted that successful urban regeneration projects are the cases which national or federal funds are flexibly utilized in response to local needs on the basis of bottom-up initiatives, as seen in the Euromediteranee project in Marseille.

16.1.4 Formulation of Strategic Spatial Planning at the City Region Scale

In Europe, rather strong public-sector intervention in land development is widely accepted in comparison with other parts of the world. Regional development activities in the long term generally accompany changes in land use. It seems rational to coordinate regional development activities through spatial planning with workable land use regulations in the European context.

Among Asian cases, Korea enacted the National Land Use and Planning Law in 2002, which merged the National Land Use Law and the City Planning Law and significantly strengthen public intervention in land development in the whole country through a consistent plan-led system. On the other hand, case cities such as Chiangmai, Yogyakarta, and Japanese cities lack institutional spatial planning measures at the regional scale, and it seems rather difficult to guide balanced development in suburban as well urban fringe areas.

In the United States, Portland a notable case in the sense that it established regional land use regulatory framework. Many other cities such as Phoenix have also recently adopted the smart growth principles in order to pursue sustainable development and to create livable cities, though the effectiveness of policies varies significantly among states and cities.

16.2 City Region and Governance

16.2.1 The Space of City Region

How can the city region be grasped spatially? In Chapter 1, the city region is defined conceptually as the space of spatial proximity to lead region-specific innovation through close face–to-face human interaction. In every regional planning process, the question of boundaries is one of the first to present itself. Unlike cities or other well-established governmental institutions, regions frequently do not have clear boundaries backed up by powers delegated from more senior levels of government or the force of law. This forces regional planning to begin not just with an apparent challenge or problem, but with the creation of findings in support of the nature of the territory to be considered by the planning itself.

In one sense, this could be viewed as an advantage for regional planning. After all, boundary conditions are dynamic, and regional planning, unlike city planning, is not stuck with obsolete or unresponsive boundaries. Consider the derivation of metropolitan areas by the US Census. The history has been that every 10 years those boundaries change due to changes in commuting and employment patterns. This serves as a graphic example of the ways in which regional boundaries change, and the advantages to regional planning of being able to define boundaries based on the nature of the problem and the conditions at the time.

However, the reality is that better bounded and more traditional governments jealously guard their home-rule prerogatives, even when they aren't in their best interest. Further, unlike bounded entities, which can and frequently do plan by looking inward, regions are best understood not as separable units but as parts of and contributors to a 'whole,' be it a state, nation, continent, or global set of relationships. Fundamentally, regions have fuzzy edges and clear centers, the opposite of the conditions encountered in the process of city planning. The challenge for regional planners amounts to enlisting bounded entities in a collaborative enterprise so that planning can proceed as if boundaries didn't matter.

In looking across the case studies, some of the city regions are clearly polycentric (Stuttgart), and others exhibit a combination of polycentric and monocentric conditions. Interestingly, none of the regions examined has stated a conscious desire to become monocentric. Quite to the contrary, those regions with a regional plan have all adopted some form of polycentric goal for their future. This may simply be a reflection of the fact that enforcing a monocentric city in regional, globalized economies is simply beyond the reach of even the most powerful central city. Further, the movement of both jobs and people to suburban and exurban areas has significantly eroded the political and economic power of central cities.

Consequently, the boundary issue, the definition of the space of regions and the territory for regional planning, will never go away. In an age of global city regions and megacities this raises another issue: when does being big matter? Is there any significance in the growing number of cities of over five million inhabitants? No matter how big a city gets, does the scale of daily life for households change, or is size notable simply because it is something we can measure and compare? Whether megacities signal a new set of concerns regarding boundaries, competitiveness, or other issues associated with the urbanization of the world's population and sustainable urban regeneration remains to be seen. Alternatively, the size of cities may simply mean that the same difficult questions confronting regional planning have now assumed a new scale, though perhaps not a new significance or challenge.

Generally speaking, this space of close spatial proximity can be realized in the range of more or less 50 km diameter where one-day business trips are feasible, though, of course, the actual coverage of city regions will change according to the level of the provision of transportation infrastructure as well as historical, social and political situations, etc. Indeed, among the case cities, those cities which have established an institutional framework for regional governance, such as Yogyakarta, Dalian, Stuttgart, and Portland, cover more or less this scale of spatial domains as their regions for regional governance. As for spatial composition of city regions, two types of city regions are identified: monocentric and polycentric. Monocentric city regions are composed of one central city and its hinterland, and polycentric city regions are composed of more than one central city and their hinterlands. Among the case cities, those areas where more than one independent city exists in spatial proximity, such as Kanazawa/Toyama, Daegu/Gumi/Gyeongju as well as Marseille/Aix-en-Provence/Fos, can be understood as polycentric. In this regard, the case of the Øresund region is particularly interesting in the sense that the attempt to create a polycentric city region is being made across national borders. Copenhagen (Denmark) and Malmö (Sweden) and their hinterlands were connected by the newly constructed cross-border Øresund Bridge and local governments of both sides started cooperation in regional economic development at the political level.

16.2.2 Local Governance

Taking two planning issues into account --- planning style (plan-led or project-led) and the extent of involvement of different stakeholders -- public decision-making styles can be classified as in Fig. 16-2. Among the styles of public decision making positioned in the figure, corporatism has traditionally been the case in European countries, which means a style of public decision making through negotiations among a few influential bodies such as government, businesses and labor unions. Pluralism seems rather characteristic of North American countries, which have wide support for the legitimacy of public voting, the participation of as wide range of the population as possible, as well market-based resource allocation. Asian

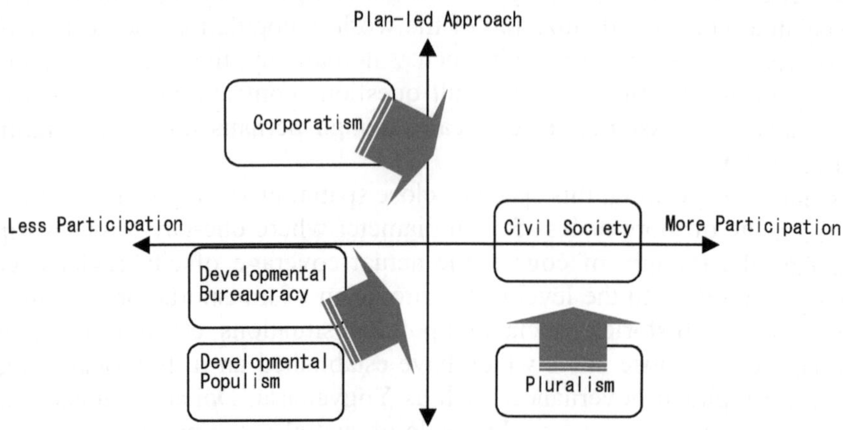

Fig. 16-2. Conceptual framework of the style of governance and planning approach for urban regeneration

countries traditionally seem to fall in the categories of either the developmental bureaucracy or the developmental populism. The developmental bureaucracy is characterized by public decision-making under established bureaucracy with strong hierarchy between the national government and the local governments, while the developmental populism is the style of public decision making under the strong influence of charismatic political leaders.

As discussed in all cases, the shift from the conventional government to governance in the style of public decision-making is observed and the partnership among various stakeholders particularly between the government sector and the private sector are being strengthened. Though the involvement of various stakeholders and the collaboration among them are understood essential to realize sustainable urban regeneration, there is also the danger that the decentralization and informalization of decision-making process in particular in large-scale urban regeneration and development projects tend to make the decision-making process blur. The transparent coordination in the arena of formulation of spatial strategy through the participation of various stakeholders is becoming an important role of spatial planning.

16.2.3 Regional Governance

The extent of the formation of regional governance varies significantly from city to city. Institutionalized regional government is not a common style for regional governance. Dalian is categorized as a single regional government type. The jurisdiction of the City of Dalian covers quite a wide area (12,574 km^2) according to the administrative structure in China. In this category of city region, public decision making at the city region scale seems rather easy in comparison with other cases. Yet, conversely, the facilitation of bottom-up initiatives of various stakeholders at the city-region scale may face difficulty because of the concentration of decision-making powers in the regional governments.

Stuttgart and Portland fall to the strong association type. They created the regional association organizations with elected councils (Stuttgart Regional Association, and the Metro, respectively) as the bodies coordinating municipalities. It seems that this type of regional governance facilitate the participation of various stakeholders because of the balanced nature of powers among various levels of government, the private sector, and the voluntary sector. In addition, it is pointed out that the Stuttgart Regional Association has the integrated functions of regional economic development policy and regional spatial development planning. The Stuttgart region is a good model on the institutional framework for coordinating regional development activities through spatial planning.

Several case cities have a form of weak association for city-region areas. The metropolitan Daegu region plan (spatial plan) was formulated in 2005 under the new National Land Use and Planning Law 2002 for the area of Daegu and neighboring municipalities where the greenbelt for Daegu is designated, though this area is much smaller than the city-region scale. Daegu city had the status of metropolitan city in 1981, which has the same status as the province and is administratively separated from Gyeongbuk province; thus, cooperation between the city and province at the city-region scale has been difficult. Recently, the mayor of Daegu and the governor of Gyeongbuk province officially agreed to cooperate in regional economic development. Several initiatives by the private sector for regional economic development have also started recently in response to Daegu city's economic difficulties due to the impacts of globalization (the textile industry, Daegu's traditional main industy, has experienced severe decline due to competition with China's emerging textile industry). The case of Daegu, where various levels of cooperation in the regional scale are emerging, can be categorized as the weak-association type. In the case of Yogyakarta, Yogyakarta Province is delineated based on the old Kingdom of Yogyakarta, which was formed as the domain of the hinterland of capital city, Yogyakarta. The province of Yogyakarta is said to have a strong historical tradition of cooperation due to its history as an old kingdom. In Yogyakarta, recent decentralization and the strengthening of municipalities seem to change such traditional conditions and facilitate the initiatives of municipalities and other private, community and citizen organizations.

Many of case cities fall into 'fragmented' type in terms of regional governance. Yet, various attempts towards institutionalizing cooperation among different sectors at the city region scale are being made in all cases. For example, Chiangmai was designated as the central city for the northern region of Thailand called the Lanna cluster in the 9th National Social and Economic Development Plan (2001-2006) and large-scale development projects such as the improvement of the Chinagmai International Airport and international cultural and sports events have been implemented under the initiative of the national government in collaboration with the private sector. Economic cooperation with neighboring countries such as Yunnan Province in China and Lao PDR is also being promoted, particularly in the field of tourism development. In Japan, under the new National Land Sustainability Plan Law of 2005, regional planning associations composed of ministries' regional branches, local governments and the private sector organizations are to be established.

In general, the bottom-up style of the formation of regional association will be difficult in polycentric city regions. Thus, the role of the national government as a facilitator will become important. Yet, in order for the national government to act as a facilitator, the national government should also create an integrated vision for regional development with close considerations of the specific needs of the region. Institutional set-up of agencies like regional offices, which are independent from line ministries and have their own regional development funds (not just functioning only as a coordinator among line ministries without their own funds), could possibly be one of the answers to this issue. Yet, there is a danger that a few powerful private-sector organizations can influence the public decision-making process due to their financial and organizational capacity in case of regional associations established in a top-down process. Thus, it is important to devise an institutional framework to support capacity building of voluntary organizations and civil societies which have sufficient financial and organizational capacity at the regional scale.

16.2.4 Governance of City Region

Regardless of the size or scale of regions, the central challenge to regional planning is finding the capacity to work effectively across, rather than within, boundaries. Much has been made of the failure to develop regional forms of government within which regional planning can take place. In the absence of a sustained and effective 'movement' to establish regional governments for regional planning, many observers and practitioners have turned to the notion of regional governance, government-like activity carried out by relatively informal associations of jurisdictions and interests. Though this form of governance as a substitute for government seems more likely in complex, multijurisdictional regional cities, it has little to show in the way of results.

Perhaps the challenge today is to turn from viewing government or governance as a means for exercising control, and instead to understand the government/governance issue at the city-region scale to be one of accountability. Regional planning can't succeed unless an environment of mutual accountability has been created and sustained. Ultimately, without some sense of regional citizenship, enforced either by a regional institution or a set of effectively mandatory reciprocal expectations, jurisdictions and interests will never temper local desires according to regional plans and needs. Fundamentally, the benefits of regional affiliations that make individual interests accountable to the whole must be tangible and compelling.

However, regional accountability, though an important cornerstone for regional governance, must be accompanied by the empowerment of citizens and communities if the outcomes are to be equitable and fair. Regional governance, like local government, only works well if affected citizens care and contribute. No citizens will care much about regional outcomes if they feel disempowered and incapable of making changes in their local community. To institutionalize boundary-crossing behavior rather than boundary-managing defensiveness, regional interests must seek and sustain local empowerment. To successfully create a culture of collaboration, individual interests need to be secure enough to imagine alternative relationships with the world around them. In the end, regional planning must, like all planning, make what is desired easy to do, and what is not desired hard to accomplish. Outcomes matter, and outcomes desired by communities and expressed in plans must be the aim of plans and planners from the outset. Creating a regional plan with no hope for implementation simply gives comfort where there should be none.

Both people and political structures are fallible. Corruption, discrimination, and age-old social and cultural conflict do not vanish before the gaze of even the most committed planner or visionary leader. As has been said in the past, 'we get the government we deserve.' We must question whether acting as if corruption and discrimination don't exist or can be countered is really useful. How should planning anticipate this, if it should at all?

16.3 Placemaking

The experiences of case cities show that for the success of urban regeneration, overarching urban development concepts (guidelines, mission statements) are indispensable. In order to actively plan for the future, cities need an overarching vision of urban development to which strategies and policies can be aligned. The main objectives of almost all strategies for urban regeneration projects are promoting mixed usage and re-urbanisation. Often, housing projects have a leading function and are offered in various forms. In many cases, temporary uses are also no longer regarded as a disturbance or handicap, but rather embraced as cultural pioneers that are essential for reintegrating previously 'forbidden or forgotten' areas. They create new images of the area.

There are evidences that cultural factors play an important role in urban regeneration. First, urban forms are directly affected by the behavior of residents in using urban space. A blend between formal, orderly urban structures and traditional, unstructured urban forms can be found in cities, such as Yogyakarta, whose residents live in a culture of duality. Duality can be understood as a status of transition where an individual or society undergoes a process of change from traditional, natural, unstructured values and behaviors toward modern, artificial, and well-structured ones. It is manifested in social behaviors as well as physical settings. Such a mixture, which may also be found in dualistic cultures of developing countries elsewhere, is not necessarily a shortcoming that needs to be amended. Indeed, if handled correctly it can be an attractive feature for visitors wishing to experience traditional living. Second, however, the same cultural trait of duality, may lead to lax behaviors with respect to the observance and enforcement of public rules. Regulating behaviors in spatial development often becomes a daunting task not only because local authorities are not sufficiently equipping themselves with good capacity for oversight and enforcement, but also because the citizens tend to revert to traditional practices that may not be in keeping with formal regulations. Thus, public regulations that are meant to enhance sustainability cannot be well implemented. It is clear from experiences in cities in the developing countries, such as Yogyakarta, that the effect of culture can have dual and often opposing consequences, i.e. attractiveness that may lead to sustainability as well as unstructured behaviors that may pose threats to sustainability. It should be a challenge for urban governments and concerned communities to strive for ways to enhance the attractive effects of culture and to minimize threats to sustainability of urban regeneration.

People in even the largest city regions live their lives in territories of their own devising. Simply put, despite the inertia of local governmental boundaries, few among us actually live our lives completely within the boundaries of a single entity. Ultimately the true scale of modern urban life is bigger than the neighborhood but smaller than the region. Yet, it is the places that exist at the human scale that define the allegiances that citizens have and develop. The implications for sustainable urban regeneration are clear: sustainable urban regeneration is accomplished as a balance between very local, human-scale activities to create great and memorable places, and large-scale, regional interventions to manage growth and address environmental and urban-system concerns. Those pursuing sustainable urban regeneration can't simply seek to be effective either regionally or locally. Both kinds of success are needed, and planners and others committed to this work must seek success both with the fine-grain task of making great places, and the coarser and less well-defined task of creating sustainable regions.

For sustainable urban regeneration to be realized, planning and redevelopment efforts must operate at both the regional and local levels, and at the landscape and community scales. Regional and local activities need to complement each other and be systematically linked. Ultimately, competitiveness has much to do with the creation and maintenance of great places, and great places result, in part, from understanding the human experience in a regional context.

MIX
Papier aus verantwortungsvollen Quellen
Paper from responsible sources
FSC® C105338

If you have any concerns about our products,
you can contact us on
ProductSafety@springernature.com

In case Publisher is established outside the EU,
the EU authorized representative is:
Springer Nature Customer Service Center GmbH
Europaplatz 3, 69115 Heidelberg, Germany

Printed by Libri Plureos GmbH
in Hamburg, Germany